Jump Start Your
Process Approach

An Indispensable Tool for Organizations
that Want To Improve Using
ISO 9001:2000, AS9100 or
ISO/TS 16949:2002

Alan J. Peterson

With technical contributions from
The Boeing Company and General Motors Corporation

QSU
Publishing Company

Acknowledgements

The publisher wishes to recognize the invaluable contribution of the following individuals without whose contribution this book would not have been possible.

The Boeing Company: Steve Malnack, Kitty Samaneigo, Danette Harris, Cyndi Koopman, Tony Marino, Tim Lee, Craig Dorman.

General Motors Corporation: Joe Bransky.

DaimlerChrysler Corporation: Hank Gryn.

Plexus Corporation: Jodi Shorma, Carl Garrett, Jim Collins, Jose Dominguez, Jay Fang, Soon-Young Won, Steve Suzuki, Antonino Cianco, Reg Shaughnessy, Wally Wegner, John Yerly, Alex Chong.

Plexcan Corporation: Peter Lane, Brian Caughlin.

Table of Contents

Introduction to the Process Approach

Many of today's most important management system standards — ISO 9001:2000, ISO/TS 16949, AS9100, TL 9000 and others — are based on a process approach. This is nothing less (and nothing more) than a system for managing your business based around a recognition that every manufactured part, component, service or software program is designed, produced, delivered and serviced through a series of interlinked processes. These processes flow from one to another in a giant loop that includes customer requirements and feedback, not unlike the network of pipes that carries fresh water to our homes each day and then on to a treatment facility.

Without a proper understanding of the process approach, organizations implementing these standards will be doomed to a poor return on investment and systemwide inefficiencies. These companies may get a certificate to hang on the wall, but they will not add value to their organizations. For little more than the cost of this book and a few hours of your time, you can learn how to implement the process approach in a manner that will add value by eliminating waste and ultimately improving your products and services.

This book is written for the legions of auditors, line operators, managers and CEOs to whom a management system registration certificate simply will not suffice to put their children through college, buy houses for their families and even pay for an occasional vacation — in other words, this book is written for organizations that want to improve.

While parts of this book may be too technical for some, others may find them elementary. It is imperative that everyone touched by your organization's network of processes be grounded with a common understanding of the process approach.

We have tried to incorporate as much implementation guidance as possible without being overly prescriptive. It is up to you to determine what works best for your organization, group, process, etc.

While it may be tempting to take a cookie-cutter approach to quality management, be forewarned that such approaches simply don't work and are a waste of resources. Quality management systems that work and provide the expected benefits germinate from an organization of people who engage one another in planned and consistent ways and who constantly reflect on the quality of that engagement and the quality of the byproduct of that engagement. This book attempts to provide you with the right tools, ideas, sequences, processes, examples, terminology and applications to get the job done.

Check Yourself

Let's start by separating the process approach implementation into just five steps. The accompanying graphic (Figure 1) will familiarize you with the tools and models discussed. It is not intended for you to be able to read every word. The graphic will be explained in due time and is repeated at the end of the book, hopefully when you have a fuller understanding of it.

Necessary Unconventionality

Let's begin our journey into the process model by reading the following phrase aloud: "quality management system implementation." I'd be amazed if anyone could come up with a better string of words to confuse and confound.

Conventional wisdom dictates that a book such as this advocate the virtues of a particular subject, not highlight its flaws. But this is not going to be a conventional guidance book, and indeed, conventional wisdom is the problem, not the solution, particularly in the case of the process approach.

Not much can be done about the murkiness of the words above and the mental images they conjure up, so this book will instead focus on practicality and attempt to convey the human side of implementation. The goal, of course, is to make the message as straightforward as possible. You will then be able to judge the merits of quality management system implementation for yourself.

Figure 1 — Five Steps to Process Approach Implementation

Step 1

To analyze the organization's system "find the processes"

Components: 1) Entire organization,
2) Process approach model (PA model).
Actions: Identify all of the main processes
without classifying them.
Deliverables: Listing of all main processes.

Step 2

To define the system's model based on work from Step 1

Components: A single concept model, in this case the Octopus, which is based on the
Customer-Oriented Process (COP). *Note: two other models are shown, the value stream
model and the Advanced Product Quality Planning (APQP) model.*
Actions: Determine model.
Deliverables: A model that identifies/classifies key processes, sequence and interaction.

Octopus model   *APQP model*

Value stream model

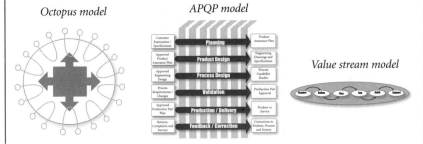

Step 3

To define the support and management processes

Components: 1) The actual processes, 2) PA model, 3) Levels.
Actions: Identify the support and management processes.
Deliverables: Listing of all support and management processes to the detail appropriate to
the organization.

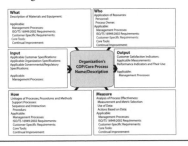

Figure 1 — Five Steps to Process Approach Implementation *(continued)*

Step 4

To adequately define and communicate the QMS

<u>Components:</u> 1) System documentation, 2) PA model.
<u>Actions:</u> Develop the documentation that can adequately communicate the system.
<u>Deliverables:</u> Documents, tools, procedures and forms.

Quality Manual

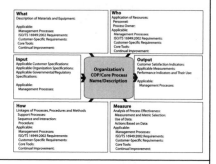

Step 5

To align first- and third-party audits by standardizing the analytic and communication tools

<u>Components:</u> 1) PA model for system, 2) PA model for internal audit, 3) PA model for third-party audit.
<u>Actions:</u> Communicate system to internal and external auditors.
<u>Deliverables:</u> 1) Picture of the system, 2) Internal audit, 3) External audit, 4) Accurate assessment of organization.

Why this approach?

There are essentially two reasons, both of them central to the mission of this guidance book.

First, most guidance to quality management system implementation tends to involve a kind of "power by association." But many guides (both verbal and written) are based on a misinterpretation of the intent of the underlying standards, requirements or specifications. This often results in a bad implementation.

Much of the so-called professional advice available in the marketplace today is just plain wrong. Here are some of the red flag phrases you might hear from one of these pseudo experts:

"The standard says … I know that because I am associated with an organization that makes its business knowing what the standard says; therefore, you must implement the standard this way."

"I know this because I know the standard; consequently, I know that you can only satisfy its requirements this way."

Of course these red flag phrases are often followed up by a well meaning, "No matter, I will make it easy for you; I will tell you exactly what you need to do" statement.

Quality management system implementation ought to be about giving an organization back to the people who work in it. After all, people are what make or break an organization, not standards and the like. The more knowledgeable and informed the people, the better able they are to use their knowledge and wisdom collectively, creating opportunities and success.

Next is a nuts and bolts explanation of process approach implementation.

Read It for Yourself

Here is one of the few absolutes in this whole process: All implementers of ISO 9001:2000 — and related industry-specific interpretations: ISO/TS 16949:2002, AS9100, QS-9000, TL 9000, etc. — must first read the standard they are attempting to implement.

Sound like good common sense? You'd be surprised how many people who have an opinion about a particular standard have never read it or read anything credible about it. They base much of their understanding and opinions on what they have heard from other people, many of whom have never read it either.

Why don't people read standards? Most often they think they will not be able to understand them. Many people associate international with technical jargon and legalese. These, in fact, are associative conclusions that do have some truth, but not enough to prevent you from reading the documents.

Two realizations need to take place. First, you more than likely read well enough to handle a standard. True, the standard must be read carefully, but that is true of anything that is worth reading. Second, the standard was written for you. It is intended to convey a message to you. No conscious attempt has been made to obscure its meaning so that only the most knowledgeable can interpret it.

Talk and Listen

After you have carefully read the standard, you need to find a place for the new ideas in your brain and get them to stay there for easy retrieval later on when you need them. Probably the best way to do that is to have a conversation about the standard with other people who have read it. The conversation need not be structured; the main idea is to flesh out your ideas.

Many people at this stage want someone to tell them what the standard means. Maybe it's human nature, fear of change, fear of the unknown, fear of making a mistake; whatever the reason, they want an expert to tell them exactly what the standard says and what they need to do about it. Some bonehead "expert" in their past probably told them they were wrong at one time. Problem is that such an approach defeats the first purpose of the standard. That is, the standard wants you to talk about it. It was written to compel you to talk to others about it. It was written so you would have to do some soul searching in order to figure out how to apply it.

Why? Because it can't tell you exactly what you should do and exactly how you should interpret it. It can't know your system. It must compel you to talk with others in your system about your system, because chances are you all have a different understanding of the system. This is also common sense since people don't tend to see things in the same way.

The standard compels you to standardize the human management system within your organization in a logical, truthful and archetypal manner. Unless telepathy and body language have evolved dramatically over recent days, the only practical way to do this is through conversation. You must determine how members of your organization perceive actions, decisions, methods, practices and interactions so that the net actions, decisions, etc., of the organization are predictable, consistent and dependable.

You need to listen to the many "voices" of your organization, some of which are not connected to an individual. The voice of the customer, the voice of the process and the voice of data are important, but too often go unheard.

You will probably need help to converse with these voices. This is a point where you might want to bring in a facilitator or trainer who is familiar with the process approach to lead your organization, steering group and/or implementation team through the conversations and to help you draw some preliminary conclusions.

BDs or 'Big Deals'

You and the others you have engaged in conversation are probably just beginning to figure out what the standard is saying to you. You are not sure you understand all of what it says, but you have a general understanding of the message or, at least, what it seems to be telling you. You are nearly ready to take on some of the more active implementation process steps.

First, though, you need to recognize the "really big deals" of the ISO 9001:2000 standard (and the industry-specific interpretations like ISO/TS 16949). This is akin to knowing how to operate an 18-wheeler prior to loading your family and all of your belongings into a semitrailer for a cross-country move.

Actually, there are probably four big deals, or BDs as we'll call them, in the standard — each related to the other. They are: the switch to a process approach itself, reliance on management decisions versus reliance on stated rules, an emphasis on measurement and metrics to support the first two BDs and committed management review to close the loop.

If you are transitioning from an earlier version of the standard, these four are the really, really big deals. They encapsulate the changes you will need to address.

BD1 – The Process Approach

Processes have long been recognized as the lifeblood of the manufacturing system, but an understanding of all the workings of an organization as a group of interrelated and interactive processes is relatively new. This new recognition is called the process approach. ISO 9001:2000 defines the process approach in the following manner.

For an organization to function effectively, it has to identify and manage numerous linked activities. An activity using resources and managed in order to enable the transformation of inputs into outputs can be considered as a process. Often the output from one process directly forms the input to the next.

The application of a system of processes within an organization, together with the identification and interactions of these processes and their management, is referred to as the process approach.

The process model employed in ISO 9001:2000 is depicted graphically in the standard. We have produced a slightly modified version in this book, but the elements are the same.

Figure 2 — Process Model Used in ISO 9001:2000

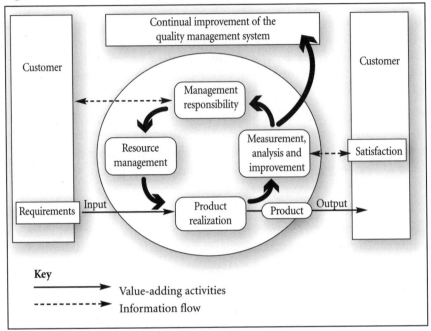

The Way a Business Operates

There was an early recognition by the leaders of industry that the process approach, if properly understood, implemented and audited, could provide substantial value to implementing organizations. Once they become familiar with the process model employed by the standard, many users have observed: "It's the way a business operates."

Though the lynchpin to improved quality system performance, the process approach is not well defined. Perhaps it has not been defined at all beyond the basic definition contained within ISO 9000 and ISO 9001:2000.

It was anticipated, though certainly not fully understood, how the process approach would be received and implemented by the many diverse cultures it would encounter. How would the practices, language and beliefs of a particular country or company be preserved within the process approach? For example, it is almost second nature for some cultures to self-audit, making the provision for internal audit superfluous and odd.

Not that those issues were totally missed by the technical committee that drafted ISO 9001:2000. They clearly understood that the process approach could not be "legislated." Measuring, monitoring, providing certain system elements, submitting pieces of evidence, etc., were parts of a quality system that could benefit from common rules. But processes were an entirely different matter. Individually, processes may have some common characteristics and common elements, but when interlinked and networked to create a complete system, their complexity, variability and uniqueness ruled out prescriptive rules. The writers of ISO 9001:2000 understood that too much regulation in this area would prove to be counterproductive.

These differences are increased by the influences of culture. In some national cultures it is second nature to check, verify and carefully attend to details, while in others it is intuitive to innovate, invent and create. To a lesser degree, the same can be said of organizational cultures. Though it seems logical and compelling on the surface that rules provide consistency, when taken too far rules create inconsistency, not consistency.

The Concept of the Process Approach

The first stumbling block in attempting to conceptualize a process approach that would meet the definition (and expectation) stated in ISO 9001:2000 "…a system of processes within an organization, together with the identification and interactions of these processes, and their management …" is to make sense of the process approach definition.

It is relatively simple to identify a process. It has an input, a transformation and an output. This can be illustrated with a simple diagram (see Figure 3).

Figure 3 — A Process

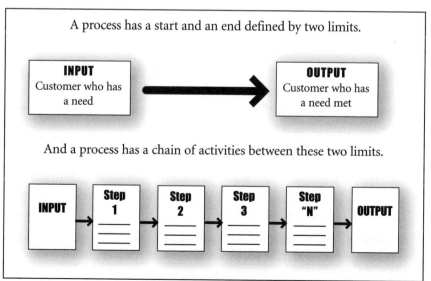

A process has a start and an end defined by two limits.

INPUT
Customer who has a need

OUTPUT
Customer who has a need met

And a process has a chain of activities between these two limits.

INPUT → Step 1 → Step 2 → Step 3 → Step "N" → OUTPUT

The definition of process approach needs to be more reflective of the process rather than a prescription for how it must be administered. The process approach therefore must be fundamental in method and generic in application. In this way, it can provide a type of benchmark that can be applied directly and/or used for comparisons.

The successful implementers of the process approach have discovered that they need to have a single, or unified, approach to processes. They have discovered it is singularly important that you have a singular model to achieve the expected benefits of a process approach and to make it possible for your organization and stakeholders to better understand how you accomplish what you do. To complement that conclusion, early implementers (based on lessons from early practitioners of the process approach — military planners, for example) are finding that a simple, fundamental question can be the start of implementation. That question: How does an organization account for all its processes in a single, unified, understandable process approach model?

Customer-Oriented/Core/Key Processes

As depicted by the writers of ISO 9001:2000 — International Organization for Standardization (ISO) Technical Committee (TC) 176 — the process model provides the first part of the answer. On the input side are customer requirements (specifications, expectations, etc.). On the output side are customer satisfaction indicators (needs met, expectations fulfilled, etc.). In the center are the organization's processes. This is a solid model as far as it goes.

The shortcoming of the graphic that is incorporated in ISO 9001:2000 is that it depicts the interaction between the customer and the organization as having a single occurrence. For most organizations, that simply is not a realistic portrayal of the number of customer interactions with the organization over the span of a typical project from concept to delivery (product realization). There are several sets of requirements and several times when the level of customer satisfaction could be affected through customer interfaces. These interfaces are referred to as customer-oriented processes, core processes or key processes (you may call them "Dan" or some other name of your choosing if you prefer). Since the standard takes a customer orientation in its process model, we have settled on customer-oriented process or COP for the purposes of this book. Several large customers such as the automotive industry also appear to be moving in this direction.

A simple definition of COPs may be those processes that begin with a customer requirement, specification, expectation or implied contract (input) and end with a fulfillment of the requirement and/or expectation being met (output).

Figure 4 — The Process Approach *(Octopus Model)*

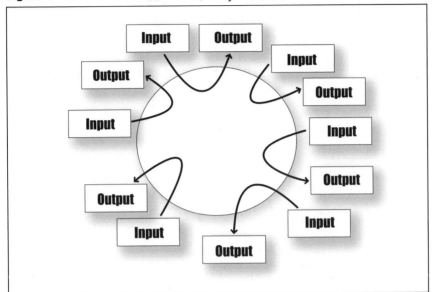

Notice, there is no specific mention about the importance of a particular process, criticality to the organization or role in all that happens within the organization. The labeling of processes as customer-oriented is simply a matter of identification. They are what they are. COPs are a straightforward and simple means with which to identify the processes that exist due to specific interfaces with the customer. They are simply a direct application of the process model with one exception: The recognition of COPs adds a critical dimension — number of interfaces. There are several interfaces between customer and organization, no longer just one. The illustration above, nicknamed the "Octopus" model, attempts to visualize the adaptation of the process model to the application of COPs.

Many COPs occurring in sequence might look like Figure 4 (conceptually). (The representation makes no attempt to portray the complexities and interconnections that are inherent in the depiction of the processes of an organization. This illustration seeks to illustrate only the external input and output interfaces between a customer and an organization as part of a COP.)

Applying the COP concept to something generic like an automobile dealership and service center might be represented like Figure 5.

Figure 5 — The Process Approach for a Car Dealership Service Center

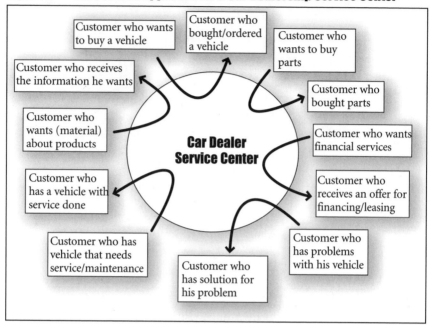

Figure 5 illustrates the central tenets behind the need to recognize multiple interactions between customer and organization. The tenets may be summarized as follows:

* Each interaction has the potential to be successful or not.
* Each interaction has the potential to lead to customer satisfaction or not.
* Each interaction has the need for a network of processes to support it.
* Each interaction must be supported by the organization, but the customer may enter into any one interaction as their first.

The plain and simple conclusion, which is central to the discussion of COPs, is this: COPs exist in every organization whether or not the organization chooses to formally recognize them in their quality or business management systems.

But are the same COPs required of all organizations? No. Of course not.

One of the first tasks in implementing a process-based quality management system is to establish a process approach that is capable of bringing about the "identification and interactions of [the] processes and their management." The underlying hope of course is that this will lead to an improved quality management system and improved performance by the organization.

Support Processes

COPs provide a direct link to customers. In other words, they are "macro" processes. But as discussed earlier, they do not represent the complete picture of what transpires within an organization. In reality, most of the activities that take place are performed by supporting processes.

A supporting process is any planned process whose input, output and transformational activities are established and maintained by an organization in order to comply with requirements or specifications and/or to meet organizational needs in supporting COPs. As we consider COPs to be macro processes, support processes are "sub-macro" processes. The relationship between COPs and support processes is not unlike the relationship of a genus and a specie in science. A genus is more general, while specie is more specific. Multiple species may be part of a single genus.

As stated earlier, COPs are more for identification rather than invention or development. The same, however, is not true of support processes. Support processes are created, organized and maintained by the organization in a manner it determines appropriate. They may be mandated by requirements, regulations, standards and the like, and some are undeniably repeated. But, by and large, the organization has significant discretion in establishing its support processes.

As a general rule, organizations need to recognize and apply two truths about support processes:

1) They are in place to support COPs or other support processes.

2) They are in place to mitigate risk.

With respect to the automotive industry, most suppliers implementing ISO/TS 16949:2002 will not be starting from scratch in terms of a quality system, so these truths have direct application in the analysis of the processes already in existence.

Although no one set of questions concerning the processes in an organization fits every situation, the following sets of questions provide a helpful foundation for establishing the need for, and adequacy of, processes.

Need

- ♦ Does the process support a COP and/or another support process?
- ♦ Does the process mitigate any current or potential risk?

Adequacy

- ♦ What is the customer requirement (input) to include both internal and external customers?

- What must be delivered to the customer (output) with respect to both internal and external customers?
- Who will be involved in the process? What are their needs with respect to training, knowledge and skills?
- What technology is needed to support the process? What equipment, infrastructure, etc., is needed?
- How will the details of activities be communicated? What instructions, procedures and methods/tools are needed?
- What will be measured and how will it be measured? What are the performance indicators?

A tool was invented several years ago that graphically depicts the interactions of processes. The tool is called the "Turtle." It was given that name due to its shape, which has a head (input), body (process definition), tail (output) and four leg categories: who, what, how and measure/analysis. A software version of this tool is now available as well (see Part III Tools, page 113).

Figure 6 — The Turtle

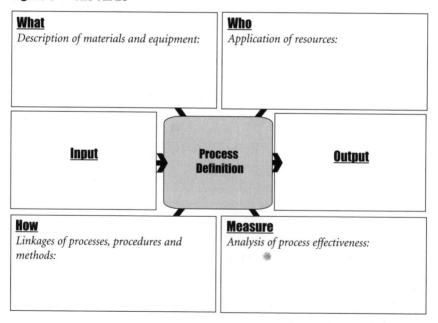

The questions with respect to need can be asked at any level of process, including the COP level. For COPs, we simply substitute the first question on our list with, Does it support a customer interface (input and output)?

At the subsupport level (support for supporting processes), the first and second questions regarding adequacy often refer to an internal customer. In such cases the final, external customer is kept in mind at all times, but the immediate need is to fulfill the need/requirement of the intended recipient of the process output.

The nearer to the COP and, of course, the COP itself, the greater the chance that a support process must be established to mitigate risk. Each process to be established or that already is in place may require other processes for support to mitigate potential risks. Each process must always competently address the questions listed earlier with respect to need and adequacy.

Points To Remember

In summary, risk is handled two ways by a management system based on the process approach structure:

1) Establishing processes to support other processes.
2) Establishing processes competently and that thoroughly answer questions regarding need and adequacy.

The bottom-line goals for a process-based system should be that:

♦ All risks are mitigated in a way that the organization and customer find acceptable.
♦ All processes are in place to mitigate risk.

Both goals cannot help but lead the organization toward effectiveness and efficiency, both of which are requirements of ISO 9001:2000. Stated simply, the ultimate product or service should work, and time, money and effort should not have been wasted in getting it to do so.

A simple, conceptual picture of a support process along with its relationship to the COP is shown in Figure 7.

Perhaps the most important element to consider when attempting to understand the process approach is what is being sought and what is being found, particularly at the conceptual level as depicted by abstract models of COPs, support processes and subsupport processes. If control is being sought, it will not be found. However, it is safe to say that control will increase to the extent that increased predictability is inherent.

What should be sought is order and visibility. Once the order, pattern, outline, model, blueprint, mold, prototype and the accompanying visibility are understood, then individuals, groups and organizations (as well as the technologies that serve them) can engage each other more effectively and efficiently.

The pattern of a process repeats itself again and again in organizations and systems. The input, transformation, output and the linkages and interconnections interface with the external environment to support the myriad of high- and low-level supporting processes.

In my experience, the same six general questions need to be asked about the processes of an organization, regardless of industry or size of business:

* What is the pattern of processes that make up my business and/or my work?

Figure 7 — Customer-Oriented Process and Support Process

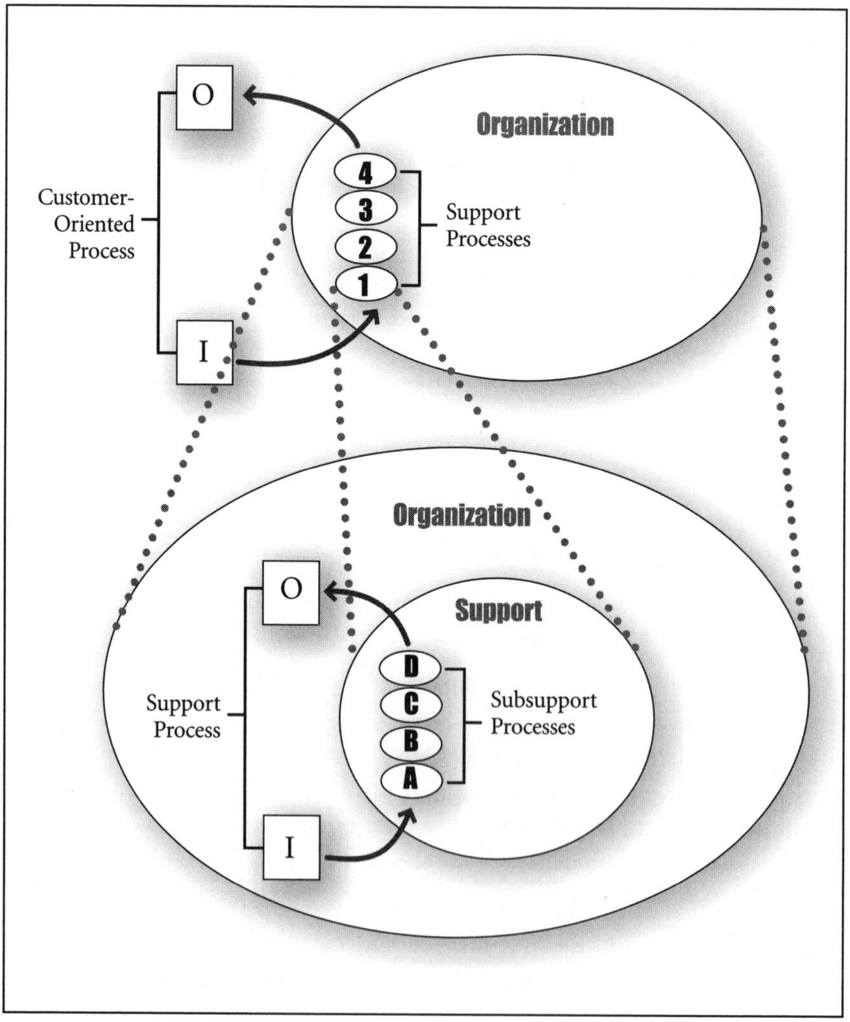

* What processes are connected?
* Why are they connected?
* Why aren't they connected?
* How are the processes connected?
* What is predictable and unpredictable about these connections?

Notice that little has been stated or asked about performance. That is intentional. The pattern and order must be understood before performance can be anticipated, measured, monitored, altered, improved and/or changed. Performance brings us to the next type of process.

This process is instrumental in assuring that improvement and change are able to take place. It is the management process.

Management Processes

Management processes are defined as those processes necessary to comply with the requirements of the standard and/or determined by the needs of an organization with respect to the role of top management. These processes focus upon the determination of the quality policy, objectives, related responsibilities and the means by which they are implemented. These are the processes that are made based on the data generated by all other COPs and supporting processes in the overall network of processes to direct and lead the organization.

The management processes along with their relationship to COPs and support processes might look like Figure 7.

Figure 8 illustrates how management processes determine standards of quality as required by the customer and/or needed by the organization. Some of these processes are planned and repeated, like management review and internal auditing. Others are unplanned but anticipated, like problem solving, mentoring, critical thinking, learning, etc.

The three categories of processes — COP, support and management — form a single/unified process approach. They are all focused on the customer. Conversely, if the processes in an organization, each in its own fashion, do not contribute in some manner to understanding what the customer requires — how that requirement can be met and making sure the requirement is met — then perhaps the process or processes are not needed.

Figure 8 — Management Processes

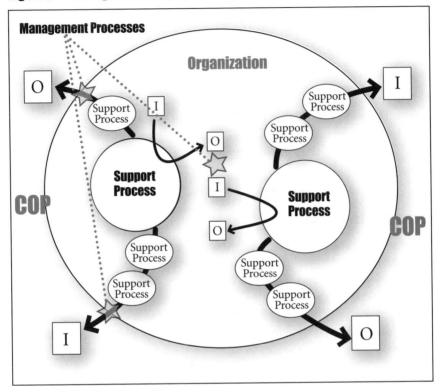

Reliance on Management Decisions Versus Reliance on Stated Rules – BD2

The 1994 version of ISO 9001 had dozens of "shall" statements. As a result, many organizations merely fulfilled the requirements for the sake of fulfilling requirements. All too often the goal became getting certified rather than creating a system that made sense for the business and would ultimately lead to improvement.

Over time, it became painfully evident to many purchasers that some implementations of ISO 9001:1994 were not focused on improvement. In fact, the quality of goods and services was for the most part no more predictable than it had been in the past.

ISO 9001:2000 attempted to make such a scenario less likely. The number of "shalls" is reduced significantly. Without the detailed prescription, each organization is forced to make some fundamental decisions about their quality management system. Fundamental and/or strategic decisions about the quality system now require management input and oversight. Upper management more clearly sees that the goal is no longer simply compliance; that implementation requires thoughtful decisions by them regarding expectations of the system driven by what is important to the business.

Essentially the focus of the standard has shifted away from the use of declarative statements by forcing companies to ask questions. Implementers are compelled to address and question several areas of their businesses. Of course we are referring to fundamental questions about the business. Previous editions of the standard also raised questions, but they were focused more on satisfying a base set of requirements. A most common theme was: Do we do this? Now an organization is expected to ask: How do we want to do this? Does this satisfy our needs and the needs of our customer? Implementers are compelled to respond to words and phrases that are less specific and more subjective — words and phrases like "sequence and interaction, effectiveness and efficiency, competent and adequate," etc. Two fundamental questions may be asked over and over throughout the standard:

1) What do these words mean in this context?

2) What does the standard expect?

Regardless of how these questions are answered, the standard is written in such a way as to compel an organization to clearly define its own system.

But defining the system only satisfies the first question. The second, What does the standard expect? is much more subtle yet much more powerful. It requires implementers to find the expectations that exist within the standard, and to be willing to develop and defend a system focused on fulfilling those expectations (as well as requirements), because the expectations are focused on processes, applications of the processes and measurement of the processes in order to support customer satisfaction.

To illustrate, parents sometimes come to the conclusion that it is often better to have fewer rules and more expectations for their children. They find that children will act more responsibly if they are given stated and implied expectations rather than only being threatened with the loss of privileges, as in the case of a rules-based system. An individual given the freedom to make choices is often more willing to subject himself or herself to a more controlling discipline than he or she would otherwise be willing to accept under a mandatory set of external rules. In the case of ISO 9001:2000, we have a set of rules granting us more freedom but expecting more accountability.

Reliance on Measurement and Metrics – BD3

To support the process approach we must create a system that is managed by situational decision making linked to companywide objectives as opposed to having a checklist of rules. The system (and each process in it) must incorporate methods to understand current performance, track problems back to their root and be capable of predicting and preventing their occurrence. This reliance on measurement and metrics is critical to the other two Big Deals already discussed.

Why?

If rules that were once based on mandating correct actions have now been supplanted by nonprescriptive expectations (elemental vs. process), then it is only natural to expect that the system will be judged on performance. Performance lends itself to measurement. Correct measurements become the primary way to truly manage a process-based system.

It has been stated that "process-based management [and thus, ISO 9001:2000] requires and encourages data-driven decision making. Without data, the process is managing you." Perhaps even more succinctly, "What gets measured, gets managed."

There are at least two issues that need to be addressed with respect to measurements. Both have a great deal to say about where measurements fail to measure up and why measurements don't provide the value they ought to.

The first has to do with "single point" measurement. The second and related issue deals with attribute and variable data points.

Single point measurement attempts to sum up a condition into one data point. An example would be an average. For the purposes of illustration let's take a look at a batting average in baseball. What does it tell you? It is the ratio of hits versus the individual's number of times at bat, minus noncounting times at bat (walks, sacrifices and hit-by-a-pitch). This number tells you something about a baseball player, but does it really tell you what you need to know? What about the number of hits the individual had with teammates on base? What about extra-base hits (all hits are equal in the eyes of a batting average)? Were these hits evenly distributed at pivotal versus nonpivotal times? Wouldn't you want to know the answers to these questions if you were paying an athlete several million dollars per year? Of course you would.

Single point measurements can be further detailed by using attribute and variable data points. Attribute data points are based on "either/or." Something is complete or it isn't. It fits or it doesn't. It is red or it isn't. Variable data points, on the other hand, attempt to combine attribute data points in order to consider how multiple factors attribute to understanding a particular condition or state. Variable data provide measurement information that lends itself to a more comprehensive view of the condition or state.

The example of a batting average leads to a discussion of other important factors impacting judgments and decisions about a particular condition or state. It shows the need for important data about the multiple factors that can impact a process, in this case the factors of determining the true value of an athlete's contribution. The owner of the baseball team has to determine, Is this athlete worth $5 million in today's market? He can base his answer on facts. The athlete's batting average for the year was .250 — nothing special. But it was .350 in the months of September and October when the team was searching for postseason glory. The athlete's slugging percentage (total bases) was .623 and his average with men on base was .474. Now this variable data gives the team owner a much better indication of the player's value than simply relying on attribute data (batting average alone).

Ensuring that the quality management system runs effectively and efficiently and that there is meaningful integration of the system within the organization's total system, requires linkage of process metrics directly to the organization's business objectives.

How an organization defines its processes, selects it models and determines metrics should ultimately be driven by these five business questions:

1. **How do I invest in my business?**
2. **How do I invest in my product or service?**
3. **How do I sell more product or service — gain more market share?**
4. **How do I fulfill existing commitments?**
5. **What do my customers think of my product?**

Measurement must consider all those determinations that the organization is compelled to quantify. A metric is the determination of which measurement or measurements are deemed to have meaning for the organization.

Committed Management Review – BD4

An organization's CEO (the top of top management) stays at the helm of the organization by being able to correctly answer these five questions over an extended period of time. The process approach, coupled with a quality management system based on ISO 9001:2000 and supported by management disciplines and practices as described by ISO 9004:2000, will provide answers to these questions. Management review is a primary mechanism for communicating the data and information produced by the processes within a system, which can then provide the foundation for answers to our questions and also serve as the primary mechanism for formulating and communicating the answers.

It is no secret that organizations often run parallel management systems to meet varying requirements, standards, regulations and the like. That is, organizations run the business by whatever means and method they like, and then they install a management system to satisfy the demands of a particular customer or regulatory agency. The CEO leads the business in such cases while the quality manager leads the quality system.

Why?

A lot of reasons can be listed here, many of them quite reasonable. There are now a lot of good reasons for the CEO to embrace the process approach and ISO 9001:2000. The reasons are better business results and meeting the obligations accepted when implementation was approved and put in motion.

Since this book is focused on the practical aspects of implementation, we will not go into too much detail on expected business results, but rather we will provide a summary of upper management obligations.

The primary obligation is to conduct management review. ISO 9001:2000 states the obligation like this: "Top management shall provide evidence of its commitment to the development and implementation of the quality management system and continually improving it effectiveness by ... conducting management reviews ..." (ISO 9001:2000 5.1 Management commitment).

The standard goes on to state: "Top management shall ensure that customer requirements are determined and are met with the aim of enhancing customer satisfaction" (ISO 9001:2000 5.2 Customer focus). "Top management shall ensure that quality objectives, including those needed to meet requirements for product are established at relevant functions and levels with the organization. The quality objectives shall be measurable and consistent with the quality policy" (ISO 9001:2000 5.4.1 Quality objectives). "Top management shall ensure that: a) the planning of the quality management system is carried out ... b) the quality of the quality management system is maintained when changes to the quality management system are planned and implemented" (ISO 9001:2000 5.4.2 Quality management system planning).

It further states: "Top management shall ensure that appropriate communication processes are established within the organization and that communication takes place regarding the effectiveness of the quality management system" (ISO 9001:2000 5.5.3 Internal communication).

Finally, top management is compelled by the standard to review the organization's quality management system, at planned intervals, to ensure its continuing suitability, adequacy and effectiveness.

"This review shall include assessing opportunities for improvement and the need for changes to the quality management system, including the quality policy and quality objectives" (ISO 9001:2000 5.6.1 Management review).

Those are the Big Deals; now we will discuss the actual steps to process approach implementation.

Step 1:
Why and What
(decision required)

Simply stated, the Why and What decision entails answering two questions:

1. Why does the organization want to implement a management system based on a process approach?

2. What is needed and/or wanted by the organization?

For most, the honest answer to the first question is: "Because we are required to do so." For a lot of companies, the decision to implement an ISO 9001:2000 based system is not a choice; it is a mandate. This often leads to answering the second question like this: "We want a certificate." Certificates for many companies represent the ticket of admission to a particular customer or industry.

Is there anything wrong with those questions and answers?

No, they are the reality. In fact, they lead us to more important questions, the central one being:

3. "Will our organization receive a return on its investment?" It's not cheap to implement and it's not cheap to maintain certification, so each organization is compelled to spend some thinking about their return on investment.

Far too many companies consider the decision a no-brainer; after all, the certificate is the price of admission. "It's a dumb thing to do, but hey, we've got to have it or we can't go to the dance," they reason.

There is a decision to be made here, and it is an important one. You must choose between having a fixed cost for your implementation with no expectation of a tangible return on investment or a variable cost with an expectation for a return on investment.

If you believe that implementation and ongoing certification to a quality management standard is a fixed cost with no tangible return on investment, then you should shop for the lowest prices on services that create the least disruption to your business and that can be completed over the shortest time. There's also no point in reading any further in this book. It will not help you. It is designed for organizations that want to improve.

There are many consultants who will be more than willing to help you spend your "expense" money on a job you have no expectations for, other than fulfilling an obligation. After all, such an assignment is practically risk free to the organization stepping in to do the work. Provide a low price and stay out of the way and you will be rewarded.

If, however, you expect a return on your investment that exceeds the initial price of admission, then read on because, as a business, you realize that expectation constitutes a prudent and realistic objective for your organization. You have taken the first steps to a meaningful implementation by expecting more than a certificate.

Step 2:
Model Development
(action required)

You may be one person in a small organization, or you could be a group in a large organization. In any case, you now need to develop a model to represent your system.

Why?

The most obvious reason is that you are doomed to fail without one, but that is more warning than explanation.

The real answer is related to the complexity of any organization. It would be impractical to discuss an entire organization and all its interconnections without some way of pulling it all together. You would quickly be inundated by the amount and depth of information. Consequently, in such situations a model becomes a useful tool. The model can be defined as a generalized, hypothetical description, often based on an analogy, used in analyzing or explaining something, particularly how the "something" works.

A model lends itself to simplifying the thing it is used to represent, but at the same time maintains those characteristics considered most critical. A model brings about at least two results. The first result is visibility. A model allows people to see the action, process or whatever. In the absence of a model, this particular action, process or whatever may not be readily apparent (like the analogy of not being able to see the forest for the trees). The second is the focus on "flow." If the model is truly going to represent the organization, then what flows through it will need to be determined. That realization, and the actions that follow, offer very real and important help to the organization. What is flowing through the system? How efficiently is it flowing?

Besides visibility and flow, the model enables organizations to understand the "sequence and interaction" (required by all ISO 9001:2000-based documents) of its management system. This understanding can lead to improvement and the support of improvement. One of the first supports is measurement. Since models support measurement, they in turn are also able to support correction, as well as improvement in effectiveness and efficiency (requirements of an ISO 9001:2000-based system).

How an organization "discovers" or develops its model depends upon several considerations. Probably the two most important are:

1. The ability of the selected model to communicate the organization's intentions.

2. The ability of the selected model to completely capture the organization's system.

Some organizations have found it effective as a first step to hold a brainstorming session of all processes supported by the organization. These processes can then be "affinitized" into groupings. Generally, "affinitized" groupings center around what may be considered to be a "core process." In other words, all processes that are somehow directly related to one another form the basis of a larger process.

The automotive industry, for example, has asked that organizations that implement ISO/TS 16949:2002, the automotive industry specification based on ISO 9001:2000, consider using the concept of "customer orientation" as their core process grouping. Thus, a COP of design validation (customer requirements = input, supplier response = output) would be a category to which the processes they identified through brainstorming could be "affinitized."

There are at least two rules that need to be kept in mind with respect to models. First, models are only effective to two levels. For example, a person might compare his or her hunger to that of a bear. "I am as hungry as a bear," they might say, but any real comparison between a human and a bear begins to lose meaning and effectiveness almost immediately. They are not covered with fur. They don't hibernate nor do they have claws. They don't generally stand in small streams, trying to catch salmon so they can devour them on the spot, though they may like sushi. In general, while a model may be applicable to multiple levels it is safer to stop after the second level.

The second rule is related to the first; it is the rule of flow. Something must flow through the model. If the original model only accounts for two levels and the system is seen as having four, then flow becomes a matter to consider. At least two models will be needed to completely explain how the system works; however, both models will need compatible "flow." They need not necessarily have the same flow, but it must be compatible. For example, the flow at a management level could be made up of information in the form of reports, contracts, memos, etc.; whereas at an operational level the flow could be materials used in the manufacturing process. Though the two flows are quite different, they are compatible when measured in terms of consistent organizational business objectives.

Let's take a look at three examples at the organizational level of workable models.

The first model is the generic Octopus model mentioned earlier.

Figure 9 — Octopus Model

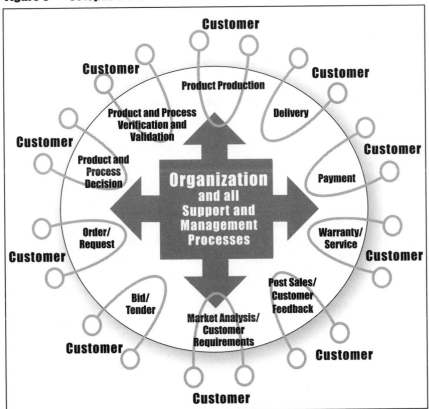

The model has as its focus what could be defined as COPs. Interfaces with the external customer set up how the organization defines the processes that drive the organization. Supporting and managing these few processes are seen as the pathway to ensuring customer satisfaction and thereby protecting and enhancing the bottom line. We will discuss the underlying processes in more detail later in this book.

Figure 10 could very well incorporate the concept of customer orientation, but approaches it from a different perspective. This approach uses the "value stream" of Lean Engineering to define its top level (some would call them core, or key, processes).

Figure 10 — Value Stream Model

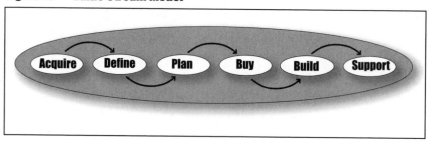

Support processes feed, and in many cases, interact with one another for each of the core/value stream processes. Management processes can be thought of as a larger circle in which these processes reside.

The third model (Figure 11) is similar to Figure 10 in that it incorporates a known model with some minor modifications and seeks to keep in place that which is already in place. The phases of Advanced Product Quality Planning (APQP) are the core processes of this approach.

Of course there are more than three models that qualify as a process approach. These three simply represent examples of acceptable ways to meet the intent, expectations and requirements of the process approach. You should develop a model for your organization that will best fit your needs, expectations and requirements.

Figure 11 — Phases of Advanced Product Quality Planning

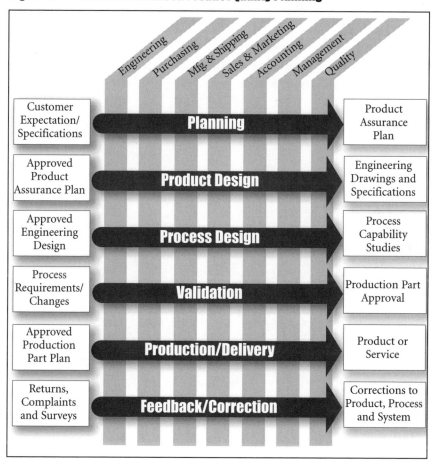

Step 3:
Management Buy In
(decision required)

The main question here is whether upper management accepts and, most importantly, supports your model.

A process approach to management has it biggest impact on management. Much is expected of management and for good reason.

An elemental approach to quality system management lends itself to compartmentalizing the quality system. Many organizations have done just that. Individuals and even whole departments become experts on certain elements of the standard. Elements are looked at as separate pieces for the sake of compliance but not as part of a larger system. The parts do not make up the whole; each part is a whole unto itself. For many organizations, the entire system only comes together during the certification audit. There are systemwide requirements, but most are handled by the same people who handle specific elements. Management doesn't really need to get involved and in many organizations they do not.

Though still made up of clauses, the overall structure of ISO 9001:2000 and the total number of requirements has changed. Its structure is based upon the type of management required for groups of activities — resources, measurement, etc. The number of requirements detailed in the standard has been significantly reduced in comparison with earlier editions.

There is now a clear mandate for the management of an organization to set the requirements since they are no longer prescribed. As a result, management must be fully engaged with the quality management system.

Why would someone in upper management want to become personally involved in the implementation of ISO 9001:2000? As stated earlier, the system can provide answers to the fundamental questions upon which they can make fact-based decisions:

1. How do I invest in my business?
2. How do I invest in my product or service?
3. How do I sell more product or service — gain more market share?
4. How do I fulfill existing commitments?
5. What do my customers think of my product?

It's only logical to presume that an organization that is able to answer these questions is fundamentally and foundationally headed in the right direction. With vigilance, improvement and success are attainable.

Step 4:
Assign Process Owners
(action required)

Each process in the (two-level) model needs an owner. This does not mean that you should replace the function of owners/leaders who already exist in the organization, nor does it mean that their functions should be scrapped. This only means that each process ultimately needs a person who has final responsibility for it. Within any organizational framework, decisions must be made that will have far-reaching impact. They need to be made by someone who has the broadest picture or with the most at stake. This is especially true in a system where process management has been adopted.

Why is this true?

Functional management/leadership tends to consider the function first and the organization second. Consequently, the management of a function like a department would, by the nature of the game, need and want to consider his function first; after all, it is his function's performance by which he is measured. A process sets up a different dynamic. The core processes, in particular, force the individuals within them to see beyond their function by involving them directly in a process that cuts across the entire organization (or at least a large portion of it).

This should not be taken as a condemnation of functional structures. They do not have to be replaced because they still have a purpose they can fulfill. They offer a sound structure for providing things that meet specific employee needs — things like payroll, counseling and professional development are only a few needs of the organization that can be efficiently and effectively supported through a functional structure.

Who should assign the process owners?

A steering committee, a management team or an implementation team are often the groups that make such an assignment. In many cases the assignment is obvious; in other cases it will take more time and often involve some trial and error. This should be accommodated in your implementation schedule.

Step 5: Process Owner Acceptance and Training
(decision required)

Is this the right process owner? That question is often best answered by the person selected. Does he or she have the right skills? Is the payoff in both intangibles and tangibles sufficient to make the commitment worthwhile? An appropriate analogy might be to imagine this is something like trying on a pair of shoes. They need to fit and also be the right kind of shoe.

You should also try to develop some overall observations of the process owners at this point. Have the key players been brought in to be part of the team? Every organization has certain individuals who can, by their influence and personalities, help or hurt an implementation. You will need to consider whether or not this situation has been addressed in the selection of the process owners.

Once the selections are made you must arrange training for the process owners' needs. The training must at minimum address these items:

- Models selected and why.
- ISO 9001:2000.
- Applications based upon the models and standard.
- Expectations of each process owner.

Step 6:
Lower-Level Definitions
(action required)

As previously stated, models are only effective to two levels. Such an assumption doesn't mean that lower-level processes may be ignored. It means that heightened visibility of the sequence and interaction of the lower-level processes adds little value at an organizational management level. Indeed, to track and maintain all individual interactions would certainly bog down the system, not increase effectiveness and efficiency.

Consequently the lower-level processes are best confined to definitions. This means the process is defined in a manner that works best for the organization. The definition should include necessary ingredients of a process — input, activities, output, measurement and the procedures and work instructions associated with each process. However, as previously stated, this need not include interactions and sequencing other than an acknowledgement of what the process takes in and what it puts out.

Many times current procedures and work instructions are used to not only provide information, but also define the scope of the process.

Once developed and accepted, these procedures and work instructions become the references listed in the first- and second-level processes, which describe sequence and interaction. A practical application would be a listing of all the lower-level process definitions (documents) that are within a second-level process.

For example, let's take a look at one of the core processes from the value stream model described earlier — the buy process (see Figure 12). To be con-

Figure 12 — Value Stream Mapping

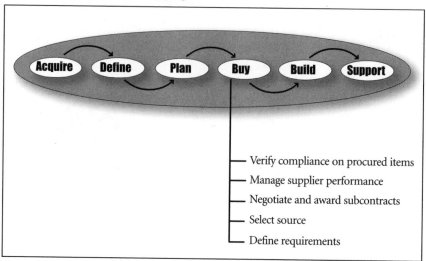

sistent, the process definitions at each level are established using the tool developed for this particular system — in this case the SIPOC (Supplier, Input, Process, Output, Customer) tool. As shown in Figures 13 and 14, the top-level model is defined. Then each of the processes of the top-level model is defined (for this example only "buy" has been defined). Then each support process/process task (depending how your organization refers to second-level processes) is defined by the SIPOC tool. These first two levels become part of the "interactive and sequenced" model. Lower-level processes (as an example, the "procedures and instructions" process step has been surrounded by a dotted line in Figure 14) can be rightly described only by the SIPOC tool. They would not, and need not, be made a part of the model, except through some indication that they exist.

It is important to understand the reasoning for not adding the sequence and interaction of the lower-level processes to the organization's process model. It is not required, as "sequence and interaction" has already been established. To do so would create such complexity in the process model as to render it less useful than in its simplified state. Finally, there really is no benefit in defining the model to a level beneath the support level.

In summary:

1. We have established the overall process model, in this case based on the value stream of Lean Manufacturing.

2. The component parts of the process model are defined via an applicable tool, in this case SIPOC, as applied to the buy process.

Figure 13 — SIPOC Tool: Process Definition Example 1

Process: Buy

Buy Process Owner:	**2**	Date: 02/17/03
Point of Contact:		Revision Date: 02/17/03

Next Higher Level Process:

Process Objective:

Support the build and support processes by acquiring parts, services and materials that are high quality, affordable and on-time.

Input (Provider) →	Process Tasks →	Output (User)
1. Prime contract (acquire) - Terms and conditions - System specifications - Delivery schedules (spares, etc.) - Program management plan - Payment plan - Work breakdown structure - Statement of work 2. Partnering agreements (acquire) 3. Part cost projections (manage programs) 4. Design package (define) - Engineering model/drawings - System specification, prime item design specifications, critical item design specifications - Process specifications - Material specifications 5. Acceptance test requirements, FTPs/ATPs and engineering change documentation (define) 6. Material requirements plan — MRP, manufacturing instruction, manufacturing bill of material (BOM), tooling plan (Plan) 7, Has Mat requirements (OSHA)	**Beginning Boundary Task:** 1. Define requirements 2. Select source 3. Negotiate and award 4. Manage supplier performance **Ending Boundary Task:** 5. Verify compliance on procured items	1. Buy-to package (Suppliers): - Design package - Manufacturing instructions - Tooling design data and plan - Purchase contract 2. Purchase contract and folio (build) 3. Receiving document (build, manage financial services) 4. Parts, subassemblies, subsystems, etc. (build) 5. Supplier performance data (manage supplier base, support) 6. Procurements specifications (support)

Process Requirements Sources:

AS9100
Prime Contract
Procedures
Government Regulations

Information Systems:

Site and process specific

Process Customer(s):

Figure 14 — SIPOC Tool: Process Definition Example 2

Process: Select Source   **4**

Buy   Process Owner: Date: 02/17/03
         Point of Contact: Revision Date: 09/12/02

Next Higher Level Process: Buy

Process Objective:

Support the buy process by finding eligible suppliers and managing the supplier approval process to ensure that acquired parts, services and materials are high quality, affordable and on-time.

Input (Provider) →	Process Tasks	→ Output (User)
1. Prime contract (acquire) - Terms and conditions - System specifications - Delivery schedules (spares, etc.) - Program management plan - Payment plan - Work breakdown structure - Statement of work 2. Partnering agreements (acquire) 3. Part cost projections (manage programs) 4. Design package (define) - Engineering model/drawings - System specification, prime item design specifications, critical item design specifications - Process specifications - Material specifications 5. Acceptance test requirements, FTPs/ATPs and engineering change documentation (define) 6. Material requirements plan — MRP, manufacturing instruction, manufacturing bill of material (BOM), tooling plan (Plan) 7, Has Mat requirements (OSHA)	**Beginning Boundary Task:** 1. Supplier submission **5** 2. Assess current approval status 3. Take warranted steps - Apply procedures and instructions 4. Manage approval process and performance **Ending Boundary Task:** 5. Verify compliance on procured Items	1. Buy-to package (Suppliers): - Design package - Manufacturing instructions - Tooling design data and plan - Purchase contract 2. Purchase contract and folio (build) 3. Receiving document (build, manage financial services) 4. Parts, subassemblies, subsystems, etc. (build) 5. Supplier performance data (manage supplier base, support) 6. Procurements specifications (support)

Process Requirements Sources:

AS9100
Prime Contract
Procedures
Government Regulations

Information Systems:

Site and process specific

Process Customer(s):

Build, support and other buy support processes

3. The second-level processes (often referred to as support processes) have been added to the overall process model.

4. The second-level/support processes have been defined via the SIPOC tool (in this case the select source support process is defined).

5. The dotted-line box designates an area where the organization would likely continue the definition process; that is, using the SIPOC tool to further define the procedures and instructions (as stated earlier, the move forward would depend upon the decisions made by the organization; that is, the organization could continue to define its processes using SIPOC, or it could hand off that responsibility to the affected functions and/or use a different model for the lower-level processes). Note: Whatever model is used must be compatible in order that inputs, outputs and measurements are compatible.

Additional, complete examples using the Octopus are provided in later chapters of this handbook as it relates to the automotive industry, aerospace and ISO 9001:2000.

Step 7: Test

(decision required)

Does the system operate as described and defined?

To continue the implementation sequence in a logical manner, it makes sense at this point to once again check the system that is under construction or reconstruction. This isn't a detailed look at this time because it is not necessary and might even become confusing. Right now you must make sure the two top-level groups of processes (at minimum) form a complete system — that each is defined to the level necessary and is "proceduralized" with established metrics.

As a practical matter, this means that a group specifically formed for this purpose, or a group already in existence, needs to perform the four checks detailed further in the following paragraphs:

Complete System — Do the top-level processes (designated in this guide as core/COP) allow a customer need to be accurately understood, transformed into a product that satisfies the customer? If not, why?

Defined — For a process to be understood by those who must be competent in it, and therefore, in the system, a definition is necessary. Often such definitions are graphic, but there is no best method. The main objective to be considered is accuracy (accuracy and detail are not synonymous, often "less is more"). Often a graphic illustration is the most accurate and most helpful way to define a process.

Proceduralized — There is no such word in *Webster's Dictionary*, but for our purposes "proceduralized" means that something can be carried out accurately over time. Often this implies an actual procedure, but not necessarily. This could involve training, making sure the technology supporting the process is correct, etc.

Metrics — How will the process be measured? Will the measurements allow the process and the system to improve over time and in a manner that ultimately helps the organization?

The answers to these questions and assessments of the pieces of the system that are already in place tend to represent an educated guess rather than science. This is as it should be. Being too precise at this time will not allow for future change. Remember: Change toward improvement is the ultimate goal.

Step 8:
Fill the Gaps and
Make the Maps
(action required)

Now that you have completed the previous steps there is a good chance that gaps will be found. After all, the process might be very new to the organization. It stands to reason that gaps will make their presence apparent.

Often the biggest gaps involve measurements. The measurement taken has nothing to do with the metric desired by the customer of the process.

Gaps can also be interpreted as steps that add no value.

Though this has probably been occurring throughout the implementation process, it is important to capture the "maps" of the process at this time. These become a kind of benchmark for the process and, by composite, the system.

While maps are all useful, it is important that your primary map be capable of serving as a model. Some forms may be too limited in their ability to represent the internal workings of your organization. The best advice is be careful with your selection.

Step 9:
Requirements Check
(decision required)

Now, it's time to see if the requirements have been met.

Even though you have crossed a large hurdle by applying the process approach to your system, all parts of the quality management system may not yet meet all requirements of the relevant standard (ISO 9001:2000, ISO/TS 16949:2002, AS9100, etc.) to which the organization is seeking compliance.

Each process must be analyzed to make sure that all applicable requirements have been met. This is also a good time to identify the requirements specific to each process. Such a record will eliminate the guesswork and investigation that is often part of the audit function in a quality management system (though the results may not be immediate). With this kind of identification, third-, second- or first-party auditors do not have to guess what the organization has applied to a particular process, and the information is readily available for examination.

Figure 15 shows a tool that can be used to identify applicable requirements, as well as other requirements, needs, etc., for the processes identified earlier. This tool can be used in conjunction with the SIPOC (Supplier, Input, Output, Customer) tool, or can be used alone. If we take the core/key process "buy" illustrated earlier and the support process, "select source," we can begin to complete the matrix. By studying the requirements (AS9100, ISO/TS 16949:2002, customer-specific, governmental, etc.) you can now see where you might have fallen short in your application of requirements (the matrix in Figure 15 is partially completed for the purposes of illustrating the completion actions).

Figure 15 — Process Approach Mapping Matrix

1	2	3	4	5	6	7	8	9
Six Process Characteristics: ○ *Has an owner* ○ *Is defined* ○ *Is documented* ○ *Linkages are established* ○ *Has records maintained* ○ *Is monitored*	*Four Support Process Questions (related to risk):* ○ *With what? (materials, equipment)* ○ *With who? (skills, training)* ○ *With what key criteria? (measurement, assessment)* ○ *How? (methods, techniques)*						Support Processes: 1) Obviously missing = *process(es) not complete* 2) Apparently missing = *performance does not meet expectation*	Requirements related to: 1) Key or 2) Support Processes 3) Management processes *not met by organization*
Key Critical Process	Support Processes for Key / Critical	Management Processes	Organization Location (Physical & Organizational)	Expected or Required Key Indicators, Measurements	Applicable Requirements	Applicable References		
(Row 1 is to be used for the key process identified in this box)	Select Source				7.4, 7.5, 7.6	PPAP		
					7.4.1, 7.4.3			

Step 10: Identify and Connect Requirements to Processes
(action required)

As a result of the work done in Step 9, the requirements can easily become a part of the procedures, maps, etc. The result is not only visibility of the processes and how they work, but also how they are managed.

It now makes sense that the audit function spends more time auditing rather than trying to figure out how things are organized. The way the system and each process works is as apparent as the map that is drawn, the narrative that has been written or the picture that has been made. The dividend is that the audit function can spend more time on value-added work.

This allows the process of auditing to focus on providing value rather than expense.

Be prepared for this step to take some time, not so much due to the complexity of the implementation at this stage but rather based on likely resistance to making a decision. You may also find that you have too few people making decisions, and you will have to make changes down the line.

Implementation Tool

The Turtle tool we discussed earlier is extremely useful in documenting your processes and system during implementation. The tool is both simple and highly visual. It will allow you to analyze each of your processes in a common visual way. Each process can be depicted graphically to include its individual parts. The Turtle tool is a good way to help you establish and maintain three central ingredients to a process-oriented system: sequence, interaction and measurement.

The following are instructions for using the Turtle, shown below (Figure 16) and described in the box diagram that follows (Figure 17). The numbers below correspond to the actual numbers on the tool.

Figure 16 — The Turtle Analytic Tool

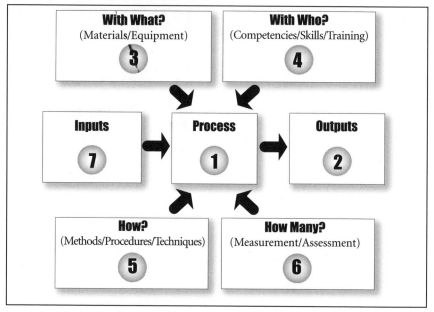

Figure 17 — The Turtle Analytic Tool

Section	Details
1	Enter COP or support process name.
2	Enter details of actual output. This may be a product or document and should be linked to actual measure of effectiveness.
3	Enter details of machine, materials (including test equipment), computer system, software used in process.
4	Enter resource requirements, pay particular attention to required skills and competence criteria, safety equipment, etc.
5	Enter details of linked process controls, support processes, procedures, methods, etc.
6	Enter the measures of process effectiveness, i.e., matrix and target.
7	Enter details of the actual input. This may be a document, materials, tooling, schedule, etc.

Using the Turtle — Process Approach Model

A more detailed form of the Turtle called the process approach model (PA model) is explained below. (The PA model is further explained later in this book.)

1) Identify the core process that you plan to analyze. Write the name of the process in the center oval.

2) Complete the information at the top of the page. The process owner should be the best person to support the completion of the tool for his or her core process.

3) Complete the "output" box to the extent possible. What does the customer expect? Which performance indicators are in place or are being developed?

4) Complete the "what" box — to the extent possible — the description of materials and equipment. What are the machines, materials, software and/or computer systems used in the process? Which management systems processes, requirements and tools apply?

5) Complete the "who" box to the extent possible — the application of resources. Who are the personnel? How have they been trained? Which management systems processes, requirements and tools apply?

6) Complete the "how" box to the extent possible — the linkages of processes, procedures and methods. The support processes listed should be those that are directly linked to the core process. These are considered to be the second level of processes. Do not extend the identification to the support of support processes. This will only lead to confusion. Try to determine the sequences and interactions. Ask yourself whether the core process interacts with other core processes. Which management systems processes, requirements and tools apply?

7) Complete the "measurement" box to the extent possible — the analysis of process effectiveness. What is measured? What type of measurement is used? What is done with the data? What are the possible actions taken as a result of the data? Which management system processes, requirements and tools apply?

8) Complete the "input" box to the extent possible. What are the specifications from the customer that drive the process forward? Don't forget regulatory and government specifications/requirements, as well as sector-specific requirements related to the input.

Process Approach Model

Organization: Completed by:

Process Owner: Date Completed:

What *(Resources)*

Description of Materials and Equipment:

<u>Applicable</u>
Management Processes:

ISO/TS 16949:2002 Requirements:

Customer-Specific Requirements:

Core Tools:

Continual Improvement:

Who *(Management Responsibility)*

Application of Resources:

Personnel:
Process Owner:

<u>Applicable</u>
Management Processes:

ISO/TS 16949:2002 Requirements:

Customer-Specific Requirements:

Core Tools:

Continual Improvement:

Input

Applicable Customer Specifications:

Applicable Customer Specifications:

Applicable Customer Specifications:

<u>Applicable</u>
Management Processes:

Organization's COP/Core Process Name/Description

Output

Customer Satisfaction Indicators:

Applicable Measurements:

Performance Indicators and Their Use:

<u>Applicable</u>
Management Processes:

How *(Product Realization)*

Linkages of Processes, Procedures and Methods:

Support Processes:
Sequence and Interaction:
Procedures:

<u>Applicable</u>
Management Processes:

ISO/TS 16949:2002 Requirements:

Customer-Specific Requirements:

Core Tools:

Measure *(Analysis and Improvement)*

Analysis of Process Effectiveness:

Measurement of Metric Selection:
Use of Data:
Actions Based on Data:

<u>Applicable</u>
Management Processes:

ISO/TS 16949:2002 Requirements:

Customer-Specific Requirements:

Core Tools:

In addition to their usefulness in identifying the COP/core processes, the completed process worksheet and records can also be used to identify support processes. They can be used as a basis for high-level management documentation as well as internal and third-party audits (see section on audits).

Note: It is not necessary to have a paper-based system. All PA model worksheets, reports and records may be maintained in an electronic format and accessed via a network.

Step 11:
The Few "Key" Measures
(action required)

The problem with measurement is that it produces data. At this point in the implementation it's time to begin paring down the measures your organization will depend upon in its decision-making processes. Ask yourself which measures tell you the most about a particular process. Which measures tell you the most about the two top-level processes? Which measures tell you the most about the system? How are the questions each business must be able to answer answered? Remember: That which is measured is improved.

Here are some additional questions to assist you:

1. How do I invest in my business?
2. How do I invest in my product or service?
3. How do I sell more product or service to gain more market share?
4. How do I fulfill existing commitments?
5. What do my customers think of my product?

The measurements taken must be sufficient, but also concise, so that management can absorb them. It does no good to make decisions based on facts, when the facts are so numerous and the total so vast that managers cannot absorb them.

The bottom line is that key measures are important because they are so useful. They will not reveal themselves easily, but they are in the system.

The most straightforward way of establishing key measures is to have "process owners" speak directly with "process customers." What is important to the customer? What does the customer need? This can be employed at every level of the management system, but it is most important at the two levels identified by your organizational process approach model. This means that the COPs/core and support processes form the scope of the "process customers," revealing to the process owners what is important to them. Since most process owners are also process customers (an output of one process is often the input of another), these actions are somewhat circular, and in a small company even more so. Nevertheless, the activity is important.

Key measurements need not only be visible to the parties they impact, they must also reflect what the customer (internal and external) needs/wants measured. Why? Only the customer truly knows his/her specific need and what will satisfy that need.

For example, we can look at this in the context of a trip to the drive-thru window of McDonalds on behalf of your family. Before you arrive you are given certain "specs." Once you get to McDonalds you relay the order to the talking electronic menu. Your order is then repeated for your review on a special screen that is part of the same electronic menu. Let's assume your order consists of the following: two hamburgers and two Cokes, one with ice and one without. You receive the order, check it, making sure there are two hamburgers and two Cokes — one with ice and one without. Everything seems to be correct so you start back home.

Once there, you present one of the hamburgers and the Coke without ice to your wife. You hand the other coke and remaining hamburger to your son. Almost immediately, your son's smile at the sight of the white bag dissolves into a grimace. "This isn't what I ordered." You review the original specifications in your mind. You were told to purchase two hamburgers and two Cokes, one without ice. Consequently, the few key measurements that seemed most important were the total number of hamburgers and number of Cokes along with the presence of ice in one the Cokes. That is what you brought home. You are reminded by your son that he wanted a plain hamburger, while the hamburgers you brought home are covered with pickles, ketchup and mustard. You, as his father, have dropped the ball in a big way since he has only eaten plain hamburgers since his potty-training years.

What was key to your son was not key to you. Your son had not conveyed to you what turned out to be a key measure to him — the type and/or absence of condiments. In this instance your son was your internal customer and he was not satisfied. Consequently you were not satisfied. You did not convey in your order that condiments, or a lack of them, was a critical measurement for you because it was a key measurement for your ultimate customer.

Step 12:
Key Business and
Customer Measures
(action required)

Once you have a handle on identifying the key measures of your processes, you need to consider how these measures fit into the larger picture prior to implementation (or once they have been implemented and are in kind of a verification phase). It's probably most helpful to separate the key measures into at least two categories: business measures and customer measures. For this purpose, the terms can be defined as follows:

> **Business Measures** — How are you meeting your obligations to the business?

> **Customer Measures** — What are the important characteristics of the outputs to your customer?

You will need to measure those items that are considered important in meeting the obligations of your business (profit margins, brand recognition, planned expansion, etc.) and those items that are considered to be important characteristics of the process outputs to your customer(s) (design validation, appearance, pricing, delivery time, etc.).

There must be some kind of balance. After all one could measure customer-mandated key characteristics to the level of customer satisfaction and still go out of business. For example, you could purchase and incorporate the latest technology in your process as one way of reducing your defective parts per million ratio, but a low return on investment for having done so could cause a complete collapse of your business.

Step 13: Integration of Management Processes
(decision required)

Check to make sure the following processes are in place:

+ Management Review.
+ Customer Satisfaction and Measures.
+ Continual Improvement.
+ Corrective and Preventive Action.
+ Records.
+ Document and Data Control.
+ Internal Audit.
+ Nonconforming Product Control.

The preceding list contains the typical processes of a business that are generally referred to as management processes. These are the processes that are generally characterized by the need for decision-making. Data is produced by each of the COP/core and support processes. The data thus produced is then processed (or should be) so that it may have meaning to those assigned to review it. Once reviewed, action can be taken based upon the available facts.

For example, management review is in place to assess the health of the quality management system and, therefore, the health of the business to the extent possible. Based upon the information conveyed in a management review, adjustments may be made and project time lines reevaluated. So, if a business and/or quality objective is not being met, proper actions can be taken to address the situation.

Step 14:
Assign Process Owners
Implementation Tasks
(action required)

It is time to assign each process owner his or her implementation task(s). One way to get your arms around this step is to consider the process owner's task list in the context of design, integration, measurement and education (DIME).

- ♦ **D** — the design of the process. In this case it could have at least two meanings. One, would be the actual design of the process: the how-it-works aspect. Second, the design refers to the way the process is communicated: the design of the visualization, graphic, narrative, etc.

- ♦ **I** — integrate the process. Make sure the process(es) interacts and integrates with those processes to which it is connected either by input or output. In a practical sense, this is accomplished by speaking with other process owners.

- ♦ **M** — measure the process. This means that a measure or measures need be put in place that make the most sense, that is, that can be accomplished, reveal the clearest picture and allow for appropriate decision making.

- ♦ **E** — educate others regarding the process. Process owners are responsible for making sure that each person in the process understands his/her role. Additionally, each person in the process needs to understand how it impacts other processes and how the process relates back to serving and satisfying the external customer, in other words a "line of sight" to the customer. There are several ways of educating process "actors," but the main emphasis should be on determining whether the delivery is effective. Functional application needs to be the focus rather than esoteric understanding. Much of this education process takes place one-on-one.

Step 15:
Process Design Review and Internal Audit
(decision required)

Like a manufacturing process, the process of implementation is served well if it is reviewed on a timely basis. The elements of this review vary according to the specific situation, but the common ones include:

* Evaluation of the design to meet requirements.
* Identification of any problems.
* Verification of the design.
* Validation of the design.
* Control of design changes.

The process design review will be accomplished (and should be) by the functions most impacted by the process and most involved in the design and development of the implementation process. The process design review is an action that occurs as the process develops.

Internal audit is another type of review. This is an evaluation by objective evaluators (those not directly involved in a particular process) who check to see if the intended process has been established correctly and whether it is effective and efficient.

Both the review and the audit are important in keeping the design and development on track.

Step 16:
Corrective Actions
(action required)

The review and audit may reveal issues that need to be resolved and corrected. The root cause of each issue must be uncovered and appropriate corrective action taken.

Two negative issues often associated with corrective action are: inaction/inadequate action and follow-up.

It is often the case that organizations are most concerned about finding problems rather than solving them. It is commonly felt to be easier and better to try to "manage" problems, rather than solve them. Consequently, corrective actions are containment actions as opposed to real fixes.

This goes hand-in-hand with another common trap: that the first fix is the right fix because the fixer wants it to be. This, of course, is also a mistake. All fixes need to be followed up. Check to see that the fix is effective. If it isn't, it isn't a fix.

Step 17: Prepare for Third-Party Audit

(decision required)

Until the auditing community has a chance to immerse itself in the study of the process approach and begins to gain experience into its application, it is necessary to keep in mind that the focus is no longer on compliance to a number of requirements detailed in the standard.

Now many of the processes, planned arrangements and documents that support them are decided and developed by the organization. This action tends to embed the requirements into the context of the processes that are identified and developed by the organization. The organization is more likely to create its own requirements/standards where none may be present in the actual standard. There is a potential for the creation of dissimilar systems that are unfamiliar to third-party auditors.

To assure that this focus does not harm the audit process, it is important that each organization consider the need for closer ties to their certification body. It stands to reason that the auditor will need to be much more familiar with the organization's processes prior to the audit than would be necessary otherwise in an elemental audit. To support that need, it is advantageous to increase the auditor's or audit team's understanding of the organizational design and development of the process approach. With that information in hand, the auditor is able to be of more value to the organization by being an objective observer who knows the paths, actions, etc., he/she must observe in order to obtain a complete picture of your organization.

Figure 18 — Process Plan for ISO 9001:2000 Implementation

#	Process Step	Type of Action	Explanation of Step	Personnel	Tool(s)	Implementation and/or Training Considerations and Notes
0	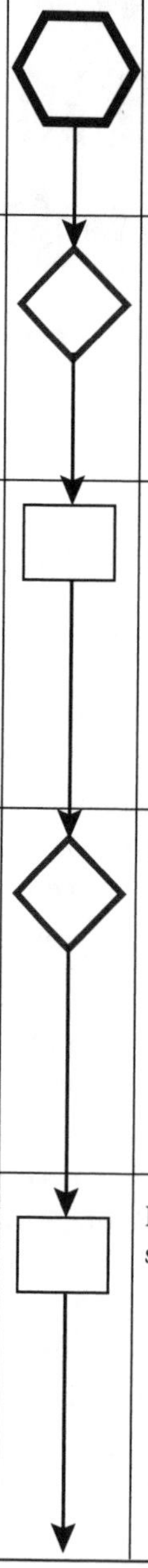	Input	An organization that needs/wants to achieve compliance/certification to ISO 9001	N/A	N/A	*Interject the linkages to the requirements at appropriate places in the training module.*
1		Decision	Does the organization wish to achieve business benefits?	Leadership	N/A	If yes, go ahead with the process described in this matrix. No, consider other options.
2		Process step	Development/ adoption/acceptance of a "two-level" model that can support process "flow" (COP/core and support)	Change agents, experts in business, thinkers	A tool to identify supplier, input, process, output, customer	This tool needs to define the supplier/input and customer/ output, and thereby defines the sequence and interaction.
3		Decision	Is the model accepted by upper management?	Leadership		Concepts that must be understood: customer satisfaction (internal and external), continual improvement
4		Process step	Assignment of process owners and definition of responsibilities which includes: design, integration, measurement and education of the process-based system	Leadership experts, process owners	A tool or strategy that allows for design, integration, measuring, education	Metrics, and the goals associated with them, are the quality (business) objectives.

Figure 18 — Process Plan for ISO 9001:2000 Implementation *(continued)*

#	Process Step	Type of Action	Explanation of Step	Personnel	Tool(s)	Implementation and/or Training Considerations and Notes
5	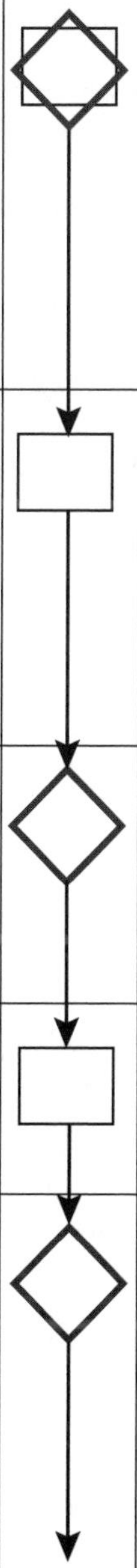	Decision and process step	(Decision) Is this the "right" process owner? Have the appropriate parties bought in? (Step) Process owner training and ISO training by leadership team (or designee)	Experts – ISO, process-based management and metrics process owners	Same as previous box	
6		Process step	Development/ adoption/ acceptance of lower-level definitions (detailed first and second levels)	Process owners, process users, process coach (expert in process management)	See steps 4 and 5 flow chart	
7		Decision	Is the process proceduralized and measured?	Process owners, process users	3-column matrix – process, procedure, metrics	Metrics, and the goals associated with them, are the quality (business) objectives
8		Process step	Map procedures and metrics to processes	Process owners, process users	3-column matrix	
9		Decision	Are the requirements of the standard met by the processes?	Process owners, ISO experts	3-column matrix — process, clause, text from clause	

Figure 18 — Process Plan for ISO 9001:2000 Implementation *(continued)*

#	Process Step	Type of Action	Explanation of Step	Personnel	Tool(s)	Implementation and/or Training Considerations and Notes
10		Process step	Identify the requirements satisfied by each process	Process owners, ISO experts	3 column matrix – process, clause, text from clause	By the actions of 9-10 you will meet the standard's requirements to identify the procedures necessary for its quality system, to map the procedures to the standard's requirements, and it identifies the relevant procedures for your people
11		Decision	What are the "few" key measures?	Process owners, metrics experts, process customers	Meeting between process owner and process customer(s)	Definitions of metrics would be helpful here
12		Process step	Identify the key business and customer measures in coordination with the customer. Business – how you are meeting your obligations to the business? Customer – what are the important characteristics of the outputs to your customer?	Process owners, metrics experts, process customers	Meeting between process owner and process customer(s)	

Figure 18 — Process Plan for ISO 9001:2000 Implementation *(continued)*

#	Process Step	Type of Action	Explanation of Step	Personnel	Tool(s)	Implementation and/or Training Considerations and Notes
13	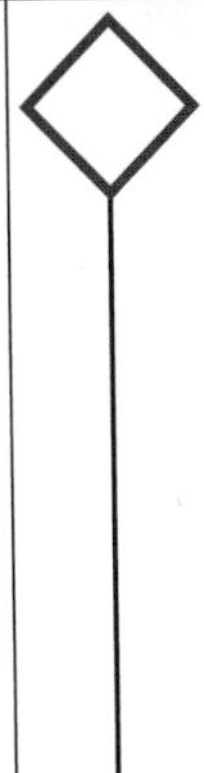	Decision	Are these processes in place? Management review, customer satisfaction and measures, continual improvement, CPA, records, document and data control, internal audit, NPC	Process owners, ISO experts	ISO 9001	
14	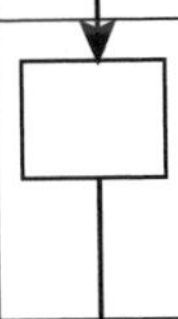	Process step	Assign process owners, DIME	Leadership experts, process owners	see steps 4 and 5	
15	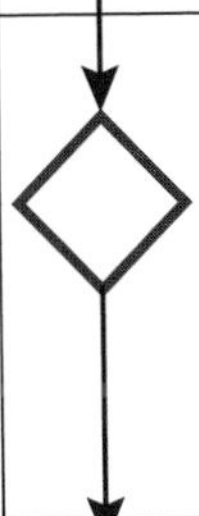	Decision	Process design review and self-assessment (internal audit)?	Auditor, experts — ISO, PBM and metrics process owners	PDR questions, website, process tools, flow charts, checklists	
16	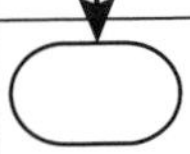	End of process	Corrective actions	Process owners	Audit findings	
17		**Output**	Prepared for a third-party audit	Third-party auditors	ISO 9001 QMS checklist	

Obviously, implementation of the process approach and subsequent auditing activities vary greatly. This book was not intended to restrict you, but rather, to give you a foundation from which to start. In that same spirit the following sections of this book are intended to give you some ideas about how you might coordinate and integrate your entire quality management system. The goal in providing this information is not to dictate how your system should look, or even how to manage it once it's in place. Rather, the purpose here is to give you some solid ideas to assist your organization down the path of continual improvement via the process approach.

Introduction to the Process Approach Model

Note: While we use ISO/TS 16949:2002 as an example, we could just as easily have used ISO 9001:2000, AS9100 or QS-9000. The focus in this example is the application of the process approach model (PA model) as an audit tool, but other applications will be evident, and the PA model's full capability will be explained and applied in the following chapter.

The PA Model

The PA model refers to a process approach management, audit and identification model and tool.

The goals of the PA model are to:

- Ensure a higher level of consistency in process approach implementation and auditing.
- Provide a meaningful record/report of the audit to the organization, the certification body (CB) and the Original Equipment Manufacturer (OEM).
- Provide a higher level of discretion for the analysis of internal, second- and third-party auditors, the CB and supplier-organization performance.

The PA model is fundamentally an extension and refinement of the "Turtle" analytic tool presently used in the automotive industry's auditor certification training for analyzing processes. Additionally, the Turtle is used in much the same way in helping to prepare quality leaders for the transition to the process approach and/or for quality leaders charged with quality system implementation.

Figure 19 — The Turtle Analytic Tool

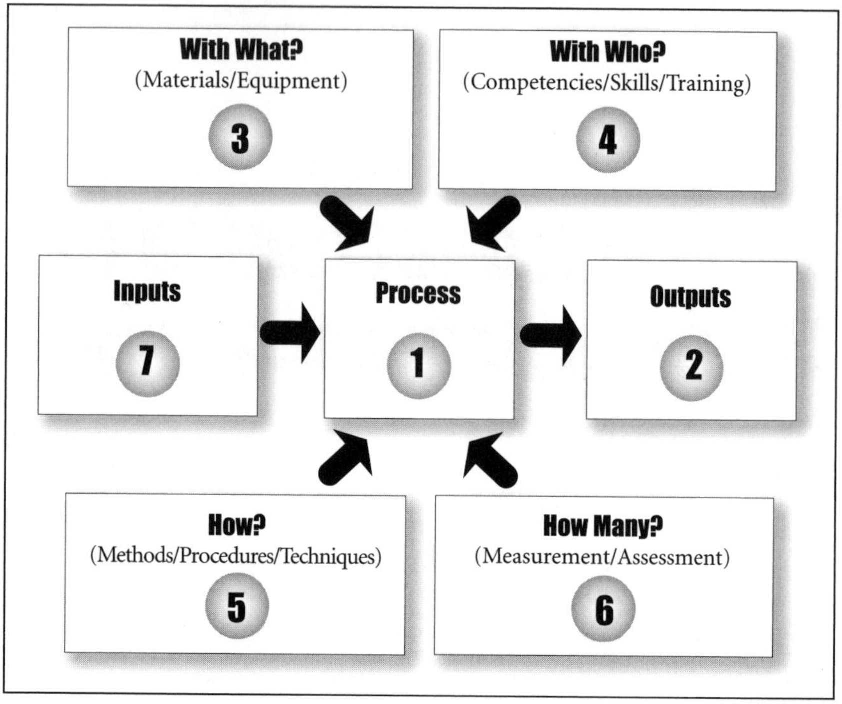

The Turtle works well in the preceding situations for a number of reasons:

* It is simple.
* It is visual, and the graphic is metaphorically accurate — it fits the process well.
* Its completion compels the individual completing it to not only under-stand the content of the process, but also the dynamics of the process.
* It is complete.

An illustration of the Turtle tool and a brief explanation of how each ele-ment is used is shown in Figure 20. It is important to understand the Turtle tool before examining the PA model.

The PA Model's Dual Function

The worksheets on pages 79 and 80 are presentations of the PA model. It simply takes the Turtle tool to the next level. The purpose is to introduce you to the "look" of the PA model. It is important to be aware of the fact that the PA model is actually two related tools/forms.

Figure 20 — The Turtle Analytic Tool

Section	Details
1	Enter COP or support process name.
2	Enter details of actual output. This may be a product or document and should be linked to actual measure of effectiveness.
3	Enter details of machine, materials (including test equipment), computer system, software used in process.
4	Enter resource requirements, pay particular attention to required skills and competence criteria, safety equipment, etc.
5	Enter details of linked process controls, support processes, procedures, methods, etc.
6	Enter the measures of process effectiveness, i.e., matrix and target.
7	Enter details of the actual input. This may be a document, materials, tooling, schedule, etc.

Applying the PA Model to Third-Party Auditing

What follows is an explanation of how the third-party/CB/registrar audit process works using the PA model, and how to apply the PA model in the audit process.

- First are illustrations of the two forms of the tool presented in their full-page versions.
- These are followed by an explanation of the tools applied to the third-party/CB/registrar audit process.
- The explanation is followed by an audit worksheet. The worksheet is provided for use by individual auditors and CBs, who may be more comfortable with a matrix-like format. The worksheet is comprised of the same basic elements as the PA model, along with additional columns for auditing decisions and notes.

Applying the PA Model to Internal/First-Party Auditing

To apply the PA model to the internal/first-party audit process really takes little or no modification. The main differences are:

- Besides the core/COP/top-level processes identified, shared and used for planning the third-party audit via the PA model, the support processes will also need to be identified, shared and used for planning the internal audit(s).

- The internal audit will be more thorough and, in general, probe deeper than the third-party audit.

- The internal audit records (PA models) should form the foundation for third-party audits and comprehensive management reviews.

Explanation of the Third-Party/CB Audit Process and the Use of the PA Model — How It Could Work

Steps

1. The PA model is sent to the organization prior to the audit. It is returned with a complete rendering of the COP/core processes (5-15 would be the typical number expected) along with the appropriate documentation for each COP/core process attached. (One worksheet should be completed for each COP/core process. Note: For future audits these would simply need to be reviewed and updated by the organization.)

2. Completion of the organizational PA model should be governed by the complexity and number of responses expected. For the categories in each box it may be possible to include titles, reference numbers, etc., that refer to an attached document or listing; in other cases it may be possible to give the complete response in the space provided.

3. The CB/registrar reviews the COP/core process PA models and from them determines: a) the audit schedule, b) the areas to focus upon and c) the need for more documentation/communication.

4. The CB/registrar may want to convert appropriate information submitted via the PA model to the auditor COP/process audit worksheet, if that would be helpful (the auditor COP/process audit worksheet uses the same categories as does the PA model, but puts them into a matrix-columned configuration, rather than the "Turtle" structure).

5. All documentation, including the worksheets if used, become the working portfolio of the third-party audit activity, and will become a part of the permanent record within the CB's files.

Process Approach Model

Organization: Completed by:
Process Owner: Date Completed:

What *(Resources)*

Description of Materials and Equipment:

Applicable
Management Processes:

ISO/TS 16949:2002 Requirements:

Customer-Specific Requirements:

Core Tools:

Continual Improvement:

Who *(Management Responsibility)*

Application of Resources:

Personnel:
Process Owner:

Applicable
Management Processes:

ISO/TS 16949:2002 Requirements:

Customer-Specific Requirements:

Core Tools:

Continual Improvement:

Input

Applicable Customer Specifications:

Applicable Customer Specifications:

Applicable Governmental/ Regulatory Specifications:

Applicable
Management Processes:

Organization's COP/Core Process Name/Description

Output

Customer Satisfaction Indicators:

Applicable Measurements:

Performance Indicators and Their Use:

Applicable
Management Processes:

How *(Product Realization)*

Linkages of Processes, Procedures and Methods:

Support Processes:
Sequence and Interaction:
Procedures:

Applicable
Management Processes:

ISO/TS 16949:2002 Requirements:

Customer-Specific Requirements:

Core Tools:

Continual Improvement:

Measure *(Analysis and Improvement)*

Analysis of Process Effectiveness:

Measurement of Metric Selection:
Use of Data:
Actions Based on Data:

Applicable
Management Processes:

ISO/TS 16949:2002 Requirements:

Customer-Specific Requirements:

Core Tools:

Continual Improvement:

Process Approach Model — If used by auditor as an audit tool

Organization: Completed by:
Process Owner: Date Completed:

What List the materials and equipment examined along with the applicable requirements, processes and tools. Record gaps, strengths and weaknesses as applicable

Description of Materials and Equipment:

Applicable
Management Processes:

ISO/TS 16949:2002 Requirements:

Customer-Specific Requirements:

Core Tools:

Continual Improvement:

Who List the resources examined along with the applicable requirements, processes and tools. Record gaps, strengths and weaknesses as applicable.

Application of Resources:
  Personnel:
  Process Owner:

Applicable
Management Processes:

ISO/TS 16949:2002 Requirements:

Customer-Specific Requirements:

Core Tools:

Continual Improvement:

Input List the specifications examined. Record gaps, strengths and weaknesses as applicable

Applicable Customer Specifications:

Applicable Customer Specifications:

Applicable Customer Specifications:

Applicable
 Management Processes:

Organization's COP/Core Process Name/Description

Output List the measurement/ indicators examined. Record gaps, strengths and weaknesses as applicable.

Customer Satisfaction Indicators:

Applicable Measurements:

Performance Indicators and Their Use:

Applicable
 Management Processes:

How List the process, procedures and methods examined along with the applicable requirements, processes and tools. Record gaps, strengths and weaknesses as applicable.

Linkages of Processes, Procedures and Methods:
  Support Processes:
  Sequence and Interaction:
  Procedures:

Applicable
 Management Processes:

ISO/TS 16949:2002 Requirements:

Customer-Specific Requirements:

Core Tools:

Continual Improvement:

Measure Record your examination of the process effectiveness. List gaps, strengths and weaknesses as applicable.

Analysis of Process Effectiveness:

  Measurement of Metric Selection:
  Use of Data:
  Actions Based on Data:

Applicable
 Management Processes:

ISO/TS 16949:2002 Requirements:

Customer-Specific Requirements:

Core Tools:

Continual Improvement:

6. The auditor(s) uses the portfolio, primarily the CB/auditor version of the PA model to conduct an on-site audit.

7. The PA model (and process audit worksheet if used) and related documentation are compared against the table or review checklist to determine if each part of the Technical Specification was adequately covered.

8. The completed CB/auditor PA model (if used), along with any nonconformance reports and worksheets/checklists, become a part of the portfolio and are used to report findings to the organization, as well as forming the basis of the auditor's report to the CB.

9. The CB makes its determinations using the complete portfolio. It notifies the organization, as well as any other required stakeholders. The portfolio becomes part of the organization's permanent file at the CB office.

All worksheets, reports and records may be on an electronic platform — this need not be a paper-based system.

COP/Core Process Audit Worksheet

Completed by: Date Completed:

Organization: Auditors:

CB/Registrar:

COP/Core Process Name/Description:

Process Element	Element Breakdown	Complete/ Gaps	Strengths	Nonconfor- mances	Comments/ Notes
Input	Applicable Customer Specifications *A listing of those items pulled from the PA model to be checked by the CB audit*	Determination regarding items listed with respect to complete- ness or gaps indicated	Determination regarding items listed with respect to strengths and opportu- nities for improvement	Determination regarding items listed as to their conformance to standards, requirements and internal system expectations *Audit reports attached*	Further explanations and/or notes to the auditor
	Applicable Organization Specifications				
	Applicable Governmental/ Regulatory Specifications				
	Management Processes				
What	Materials and Equipment				
	Management Processes				
	ISO/TS 16949 Requirements				
	Customer- Specific Requirements				
	Core Tools				

Process Element	Element Breakdown	Complete/ Gaps	Strengths	Nonconfor- mances	Comments/ Notes
What (cont'd)	Continual Improvement				
Measure	Measurement and Metric Selection				
	Use of Data				
	Actions Based on Data				
	Internal Audit Results				
	Management Processes				
	ISO/TS 16949 Requirements				
	Customer-Specific Requirements				
	Core Tools				
	Continual Improvement				
Who	Application of Resources				
	Personnel				
	Process Owner				
	Management Processes				
	ISO/TS 16949 Requirements				
	Customer-Specific Requirements				

Process Element	Element Breakdown	Complete/ Gaps	Strengths	Nonconformances	Comments/ Notes
Who (cont'd)	Core Tools				
	Continual Improvement				
How	Linkages of Processes, Procedures and Methods				
	Support Processes				
	Sequence and Interaction				
	Procedures				
	Management Processes				
	ISO/TS 16949 Requirements				
	Customer-Specific Requirements				
	Core Tools				
	Continual Improvement				
Output	Customer Satisfaction Indicators				
	Applicable Measurements				
	Performance Indicators and Their Use				
	Management Processes				

Applying the PA Model to ISO/TS 16949
(Implementation & Audit Applications)

Key Understanding

All organizations that wish to supply goods or services to the automotive OEMs must understand and live by two key concepts:

1. Line of sight.
2. Product realization.

Line of Sight

Keep the customer needs, wants, requirements and expectations, both internal and external, in view at all times. Each member of an organization must keep two customers in mind at all times, his or her immediate customer and the external customer. For suppliers to the OEMs, and for any organization in the automotive industry supply chain, OEMs are the ultimate external customers.

Product Realization

The automotive OEMs care first and foremost about the product coming out of their supplier organizations. Obviously they understand that there is more to a supplier organization than the manufacturing/processing function, but for obvious reasons, their main concerns are the products that eventually end up in their assembly plants that become part of the vehicles they sell.

The PA Model

The PA model helps supplier organizations, including CBs, keep both key concepts — line of sight and product realization — in central focus. The PA model also delivers much more. For all stakeholders in the automotive manufacturing process the PA model provides the needed visibility to all processes necessary to keep the system running smoothly. Visibility allows the supplier to understand his or her business better and to apply corrections and improvements where they will do the most good. Visibility allows the OEM (or other customers in the supply chain) to examine the processes that make up the system. It is not so they can meddle in your affairs, but so they can assist you in making improvements that can help both the OEM and you. Visibility allows the third-, second- and first-party auditors to do their jobs in an effective and efficient manner. They can spend more time examining and less time searching if the processes are visible. Visibility allows the workers in the processes and systems to understand where they fit into the rest of the processes and the system.

You cannot correct or improve what you cannot see.

Applying the PA Model To Realize Benefits

Organizations can use the PA model to map their processes in a manner that is understandable to the OEMs and to those who represent the OEMs — third- and second-party auditors. Plus there is no wasted time in doing so; that is, no other conversion, translation or delineation is necessary. Once mapped (and maintained for accuracy) the requirements of ISO/TS 16949:2002 for "sequence and interactions" are met along with nearly every other system requirement. How the system is managed and how it meets customer expectations and requirements cannot be satisfied by the PA model alone, obviously. But because the system is visible, integrated, focused on the customer (internal and external) and more amenable to planned arrangements, it is a huge step in the right direction. It cannot force management commitment, for instance, but it can make it harder to deny the obvious.

Applying the Automotive Process to the PA Model

The automotive industry has suggested that each supplier organization consider its COPs. The COPs are not mandatory or required, but rather a good starting point for an organization to analyze its functions in terms of the process approach. Identification and application of COPs set up a fairly easy process approach hierarchy to understand, implement and follow. COPs are those processes that interface with external customers. Support processes, the second level, are those processes that support or feed the COPs. Management processes are those processes that keep the COPs and support processes running smoothly.

Can an organization set up its quality management system with COPs, supporting processes and management processes? Chances are they already have; they simply have not acknowledged it. They have COPs or they would not be in business. They have some type of support processes network to support what they do with their customers. And they must make decisions, collect data, make changes, etc., all of which are actions that are required to manage their processes and their overall system. Other logical ways to think about how an organization works may be just as valid.

The system your organization defines should not only work but also be visible and complete. Be prepared to explain and sometimes defend your system to people who do business with the organization, especially if that customer is an OEM or one of its emissaries (third- and second-party auditors).

10 Customer-Oriented Processes

The 10 generic COPs suggested by the automotive OEMs are:

1. Market Analysis/Customer Requirements.
2. Bid/Tender.
3. Order/Request.
4. Product and Process Design.
5. Product and Process Verification/Validation.
6. Product Production/Manufacturing.
7. Delivery.
8. Payment.
9. Warranty/Service.
10. Post Sales/Customer Feedback.

The preceding processes need not be formally recognized by the supplier organization. They certainly can be, but the main point made by the OEMs is that these processes represent typical interfaces with them. Each, of course, may lead to customer satisfaction or dissatisfaction.

Product Realization — A Focus

Ask any OEM and they will tell you that of the 10 COPs, items four through seven are most important and number six is probably the most important of all.

If you have any doubts about this, simply examine the checks, mechanisms, tools and/or methodologies that each of the OEMs have in place to make sure the parts they receive are good parts. They tell the real story.

What does this mean to a supplier organization? It provides focus. An organization can't stop supporting the other COPs, but it can make sure that they are aligned with those that are most important to their customers.

If, for example, the supplier organization is providing the service of auditing (a registrar), then those COPs should take the highest priority. But once again, it is a matter of focus. If four through seven are audited well, the other COPs will most certainly be covered, simply because they can't be avoided. So, the message seems to be, find a part/product and follow it up and down the manufacturing stream.

COPs Mapped

In examples on the pages that follow the 10 generic COPs are mapped (though not completely — because the application is nonspecific, not all sections can be completed) via the PA model. What will be included are some suggested and generic support and management processes, as well as some completion of applicable measurements and requirements. **Keep in mind these are only suggestions.**

Following the example COPs on the PA model will be a look at the supplier's system from an OEM perspective. This is no different from what has already been discussed; it simply puts the emphasis in a more visual manner in the context of COPs.

Overall System Structure

To help you visualize the overall structure, the graphic on the following page summarizes the (1) overall model identification, (2) identification of COPs, (3) analysis of COPs with the PA model and finally (4) the recording of COPs on PA models from information obtained in the analysis, as well as internal and external requirements/analysis.

A Way To Think About It

Perhaps a good way to think about the structure and relationship of this graphic is to imagine a computer software program that allows you to build a virtual structure. The first structure is a circle, which symbolizes your organization. Loops are added as the COPs are identified and defined. Next, the PA model is used to see if the COPs have considered and mitigated all risks. As a consequence of the analysis, and to record the findings, the PA model is fully used. The PA model provides direct assistance to auditing, record keeping, identification of related processes, requirements, measurements, etc., and provides a ready means to communicate the structure and operation of the system to stakeholders. Not shown in the graphic, but which would also be a part of this virtual system, would be PA models completed for each support and management process identified in the COP/core PA model. Think of it this way: Tip the graphic on its side and support-process PA models would be connected to each COP PA model.

Figure 21 — Total Organizational System and Process Analysis

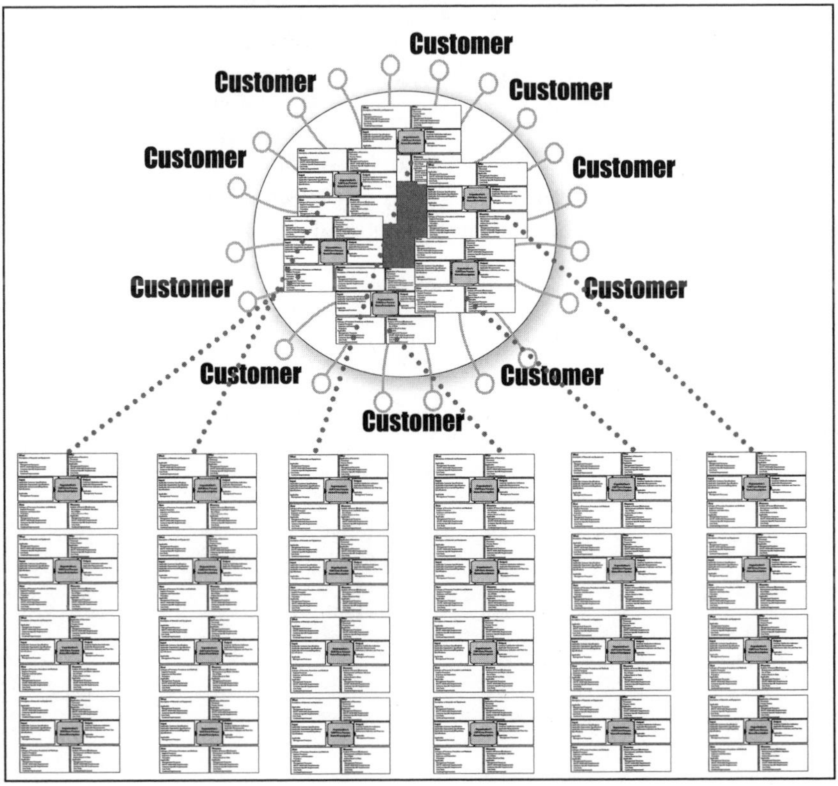

Mapping a System

The following PA model uses a generic COP identified by the automotive industry to provide an example of a completed PA model. The details of the input, output, who, what, how and measure are provided generically. The intent is to give a person using this tool some insights into what may be placed in the boxes and how that may help the organization or group to which that individual belongs. The details provided are generic; therefore, some of them will fit your situation, while others may not. Like most things related to the process approach and management system implementation, as much flexibility as possible is being provided within a framework of a generic system, so that uniqueness of the organization is embraced and that the final system will work for all members of the organization, not just auditors and quality management system personnel.

Successive PA models must be completed for COP/core-level processes and for each support process. Once completed, your quality management system (and business system) will be visible and be on its way toward a process approach management system.

Suggestion: For ease of administration and access, it will be quite helpful if the system captured in PA models is placed on an electronic platform, particularly a data-based platform.

Figure 22 presents the OEM perspective, which fully coordinates and integrates with the product realization COP. The perspective is provided to emphasize OEM expectations and to verify the utility of the PA model.

Process Approach Model

Organization: Completed by:
Process Owner: Date Completed:

What *(Resources)*

Description of Materials and Equipment:
Surveys, ads, office equipment, computers,
software, communication equipment

Applicable
Management Processes: Strategic planning,
  management review, market analysis
*ISO/TS 16949:2002 Requirements: Clauses 4
  & 5 have the most direct application*
Customer-Specific Requirements: Not often
  applicable here
Core Tools: APQP has most direct application
Continual Improvement: As applicable

Who *(Management Responsibility)*

Application of Resources: Sales staff, marketing,
top management, customer representative
  Personnel: See above
  Process Owner: Individual from sales and/or
  marketing

Applicable
Management Processes: Strategic planning,
  employment hiring, management review,
  financial management
ISO/TS 16949:2002 Requirements: Clauses 4 &
  5 have most application
Customer-Specific Requirements: Not often
  applicable
Core Tools: N/A
Continual Improvement: As applicable

Input

*Applicable Customer
Specifications:* Per the product or
  service needed

*Applicable Customer
Specifications:* Per new customer
  or current product development

*Applicable
Governmental/Regulatory
Specifications:* As applicable to
  product or service

Applicable
Management Processes: New
business decisions, strategic
planning

**Organization's
COP/Core Process
Name/Description:**
*Market Analysis/
Customer
Specifications*

Output

Customer Satisfaction Indicators:
  Quote communication/RFQ

Applicable Measurements:
  Number of quote communications and size of quote

*Performance Indicators and Their
Use:* Number and size, or
  changes to, communication

Applicable
  Management Processes:
  Customer satisfaction

How *(Product Realization)*

Linkages of Processes, Procedures and Methods:
Interface of customer need and appropriate
personnel is key
  Support Processes: Market research, advertis-
  ing, direct mailing, product benchmarking,
  sales history, customer supply list
  Sequence and Interaction: First key process
  Procedures: As applicable
Applicable
  Management Processes: Strategic planning,
  resource management
  ISO/TS 16949:2002 Requirements: Clauses 4 & 5
  Customer-Specific Requirements: Limited applica-
  tion
  Core Tools: APQP has most direct application
  Continual Improvement: As applicable

Measure *(Analysis and Improvement)*

Analysis of Process Effectiveness: Focused on
maintaining current business and success at
establishing new business
  Measurement of Metric Selection: Customer
  size/base, customer base identification, per-
  cent of market, ratings, competition data
  Use of Data: Decisions regarding resources,
  capability and capacity
  Actions Based on Data:
Applicable
  Management Processes: Management review
  ISO/TS 16949:2002 Requirements: Clause 8
  most applicable
  Customer-Specific Requirements: Often N/A
  Core Tools: SPC has most direct application
  Continual Improvement: As applicable

Figure 22 — Automotive Process Flow

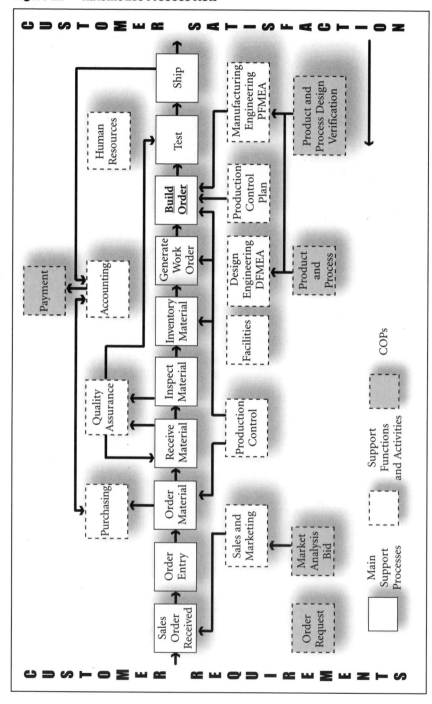

Applying the PA Model to AS9100

(Implementation & Audit Applications)

Note: While the approach described here is very similar to that of the automotive industry, there are some important differences in terminology and focus.

Key Understanding

All aerospace organizations, suppliers and OEMs/primes must understand and live by three key concepts:

1. Line of sight to the customer.
2. Product realization.
3. Oversight.

Line of Sight

Keep the customer needs, wants, requirements and expectations, both internal and external, in view at all times. Each member of an organization must keep two customers in mind at all times, his or her immediate customer and the external customer. For suppliers to the OEMs/primes, the external customers are the aerospace OEMs/primes. For the OEMs/primes, it is the people who buy the finished products.

Product Realization

The aerospace OEMs/primes care first and foremost about the product coming out of their supplier organizations. Obviously they understand that there is more to a supplier organization than the manufacturing/processing function, but for obvious reasons, their main concerns are the components and materials that eventually end up in their assembly plants that become part of the aerospace products they sell.

Oversight

The aerospace industry is a highly regulated industry. The main reason for the amount and extent of oversight is directly related to safety. The Federal Aviation Administration (FAA) is one of several regulatory bodies that serves as a kind of customer and partner. The FAA's job is to represent the public in matters of safety.

The PA Model

The PA model helps organizations, including quality management CBs, keep all three key concepts — line of sight, product realization and oversight — in central focus. The PA model also delivers much more. For all stakeholders in the aerospace manufacturing process the PA model provides the necessary visibility to all processes needed to keep the system running smoothly. Visibility allows the supplier or OEM/prime to understand its business better and to apply corrections and improvements where they will do the most good. Visibility allows the OEM/prime (or other customers in the supply chain) to examine the processes that make up the system. This is not so they can meddle in your affairs, but so they can assist you in making improvements that can help both the OEM and you. Visibility allows the third-, second- and first-party auditors to do their jobs in an effective and efficient manner. They can spend more time examining and less time searching if the processes are visible. Visibility allows the workers in the processes and systems to understand where they fit into the rest of the processes and the system.

You cannot correct or improve what you cannot see.

Applying the PA Model To Realize Benefits

Organizations can use the PA model to map their processes in a manner that is understandable to the OEMs/primes and to those who represent the

OEMs/primes — third- and second-party auditors. Plus there is no wasted time in doing so; that is, no other conversion, translation or delineation is necessary. Once mapped (and maintained for accuracy) the requirements of AS9100 for "sequence and interactions" are met along with nearly every other system requirement. How the system is managed and how it meets customer expectations and requirements cannot be satisfied by the PA model alone. But because the system is visible, integrated, focused on the customer (internal and external) and more amenable to planned arrangements, it is a huge step in the right direction. It cannot force management commitment, for instance, but it can make it harder to deny the obvious.

Aerospace Process Applied to the PA Model

The aerospace industry has adopted the process approach to quality systems management. It has been suggested that processes occur in some type of hierarchy in every organization, and that that the processes are linked and interactive. No specific hierarchy is required, but regardless of what overall approach or model is adopted, it is a good starting point for an organization to analyze its functions in terms of the process approach. Identification and application of top-level processes set up a fairly easy hierarchy to understand, implement and follow. Often top-level processes (by whatever name they are labeled) are those processes that interface with external customers. Support processes, the second level, are those processes that support or feed the top-level processes. Management processes should also be considered in order to provide processes that keep the top-level and support processes running smoothly.

Can an organization set up its quality management system with the labels and levels explained in the preceding paragraph? Chances are most organizations already have; they simply have not acknowledged it. They have top-level processes that connect directly to customers, or they would not be in business. They have some type of support process network to support what they do with their customers. And they must make decisions, collect data, make changes, etc., all of which are actions that are required to manage their processes and their overall system. Other logical ways to think about how an organization works may be just as valid.

The system your organization defines should not only work but also be visible and complete. Be prepared to explain and sometimes defend your system to people who do business with the organization, especially if that customer is an OEM/prime or one of its emissaries (third- and second-party auditors).

10 Top-Level Processes

The 10 generic top-level processes that interface with OEM/prime (that is OEM/prime as an external customer) are:

1. Market Analysis/Customer Requirements.
2. Bid/Tender.
3. Order/Request.
4. Product and Process Design.
5. Product and Process Verification/Validation.
6. Product Production/Manufacturing.
7. Delivery.
8. Payment.
9. Warranty/Service.
10. Post Sales/Customer Feedback.

The preceding processes need not be formally recognized by the supplier organization. They certainly can be, but the main point made by the OEMs is that these processes represent typical interfaces with them. Each, of course, may lead to customer satisfaction or dissatisfaction. Again, these are only examples.

Product Realization – A Focus

Any OEM/prime will tell you that items four through seven of the 10 top-level processes are most important to them and number six is probably the most important of all.

If you have any doubts about this, simply examine the checks, mechanisms, tools and/or methodologies that each of the OEMs/primes have in place to make sure the parts/products/services they receive are good parts/products/services. They tell the real story.

What does this mean to a supplier organization? It provides focus. An organization can't stop supporting the other top-level processes, but it can make sure that they are aligned with those that are most important to their customers.

If the supplier organization is providing the service of auditing (a registrar), then those top-level processes should take the highest priority. But once again, it is a matter of focus. If four through seven are audited well, the other top-level processes will most certainly be covered simply because they can't be avoided. So, the message seems to be, find a part and follow it up and down the manufacturing stream.

Top-Level Processes Mapped

In the pages that follow, an example of a generic top-level process is mapped (though not completely — because the application is nonspecific, not all sections can be completed) via the PA model. Included are some suggested and generic support and management processes, as well as some completion of applicable measurements and requirements. Keep in mind these are only suggestions.

Overall System Structure

To help you visualize the overall structure, Figure 23 (1) summarizes the overall model identification, (2) identification of top-level processes, (3) analysis of top level processes with the PA model and finally (4) the recording of top-level processes on PA models from information obtained in the analysis, as well as internal and external requirements/analysis.

A Way To Think About It

Perhaps a good way to think about the structure and relationship of this graphic is to imagine a computer software program that allows you to build a virtual structure. The first structure is a circle, which symbolizes your organization. Loops are added as the top-level processes are identified and defined. Next, the PA model is used to see if all top-level processes have been considered and all risks have been mitigated. As a consequence of the analysis, and to record the findings, the PA model is fully used. The PA model provides direct assistance to auditing, record keeping, identification of related processes, requirements, measurements, etc., and provides a ready means to communicate the structure and operation of the system to stakeholders. Not shown in the graphic, but which would also be a part of this virtual system, would be the completion of PA models for each support and management process identified in the top-level PA models (there are a total of 10 top-level PA models in this graphic to reflect the 10 COPs suggested by the automotive industry). This PA model is intended to symbolize the defined processes supporting each top-level process. The number then, is not a magic number — there could be more, or less. The graphic is provided to help you visualize the larger system and how it is connected.

Figure 23 — Total Organizational System and Process Analysis

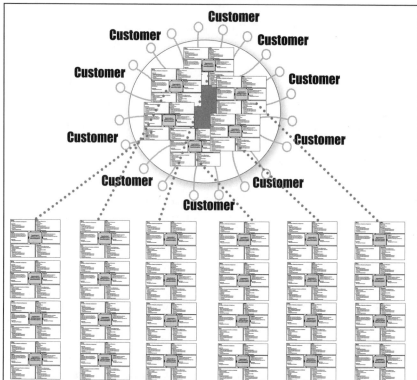

Mapping Top-Level Processes

The following PA model on page 99 uses a generic top-level process that could be used by the aerospace industry to provide an example of a completed PA model. The details of the input, output, who, what, how and measure are provided generically. The intent is to give a person using this tool some insights into what may be placed in the boxes and how that may help the organization or group to which that individual belongs. The details provided are generic; therefore, some of them will fit your situation, while others may not. Like most things related to the process approach and management system implementation, as much flexibility as possible is being provided within a framework of a generic system, in order that uniqueness of the organization is embraced and that the final system will work for all members of the organization, not just auditors and quality management system personnel.

Process Approach Model

Organization: Completed by:
Process Owner: Date Completed:

What *(Resources)*

Description of Materials and Equipment: Office equipment, computers, software, communication equipment

Applicable
Management Processes: Strategic planning, management review, market analysis
ISO/TS 16949:2002 Requirements: Clauses 4 and 5 have the most direct application
Customer-Specific Requirements: Not often applicable here
Core Tools: APQP
Continual Improvement: As applicable

Who *(Management Responsibility)*

Application of Resources: Training on bid writing, financial analysis
    Personnel: Sales, customer rep., top management (depending on the size of the bid)
    Process Owner: Sales

Applicable
Management Processes: Strategic planning, resource management, management review
ISO/TS 16949:2002 Requirements: Clauses 4 and 5 have most application
Customer-Specific Requirements: Tied to bid specs
Core Tools: APQP
Continual Improvement: As applicable

Input

Applicable Customer Specifications: Per the product or service needed

Applicable Customer Specifications: Per new, custom or current product development

Applicable Governmental/Regulatory Specifications: As applicable to product or service

Applicable
Management Processes: New business decisions, strategic planning

Organization's COP/Core Process Name/Description: *Bid/Tender*

Output

Customer Satisfaction Indicators: Quote communication/RFQ

Applicable Measurements: Number of quote communications and size of quote

Performance Indicators and Their Use: Number and size lead to affirmation of or changes to communication

Applicable
Management Processes: Customer approval/response

How *(Product Realization)*

Linkages of Processes, Procedures and Methods:

    Support Processes:
    Sequence and Interaction:
    Procedures:

Applicable
Management Processes:
ISO/TS 16949:2002 Requirements:
Customer-Specific Requirements:
Core Tools:
Continual Improvement:

Measure *(Analysis and Improvement)*

Analysis of Process Effectiveness: Landing the bid sought, customer size and base, ID customer base, customer percent of market
    Measurement of Metric Selection: Rating, competition outputs
    Use of Data: Decision — resources, capability and capacity
    Actions Based on Data: Go forward or not, steps forward

Applicable
Management Processes: Management review
ISO/TS 16949:2002 Requirements: Clause 8 has the most application
Customer-Specific Requirements: Often N/A
Core Tools: SPC
Continual Improvement: As applicable

Successive PA models must be completed for top-level processes and for each support process. Once completed, your quality management system (and business system) will be visible and on its way toward a process approach management system.

Suggestion: For ease of administration and access, it will be quite helpful if the system captured in PA models is placed on an electronic platform, particularly a data-based platform.

Applying the PA Model to ISO 9001:2000
(Implementation & Audit Applications)

Note: While the approach described here is very similar to that of the automotive and aerospace industry, there are some important differences in terminology and focus.

Key Understanding

All organizations that wish to supply goods or services to other organizations/customers must understand and live by two key concepts:

1. Line of sight.
2. Product realization.

Line of Sight

Keep the customer needs, wants, requirements and expectations, both internal and external, in view at all times. Each member of an organization must keep two customers in mind at all times, his or her immediate customer and the external customer. For suppliers to the customers, the external customers are whoever purchases their product or service.

Product Realization

Customers most often care first and foremost about the product or service produced. Obviously they understand that there is more to a supplier organization than the manufacturing/processing function, but for obvious reasons, their main concerns are the products that become part of their products and to that end impact their customers' satisfaction.

The PA Model

The PA model helps organizations, including CBs, keep both key concepts — line of sight and product realization — in central focus. The PA model also delivers much more. For all stakeholders in the supply chain the PA model provides the necessary visibility to all processes needed to keep the system running smoothly. Visibility allows the supplier and/or customer to understand his or her business better and to apply corrections and improvements where they will do the most good. Visibility allows the customer to examine the processes that make up the system. This is not so they can meddle in your organization's affairs, but so they can assist you in making improvements that can help both the customer and you. Visibility allows the third-, second- and first-party auditors to do their jobs in an effective and efficient manner. They can spend more time examining and less time searching if the processes are visible. Visibility allows the workers in the processes and systems to understand where they fit into the rest of the processes and the system.

You cannot correct or improve what you cannot see.

Applying the PA Model To Realize Benefits

Organizations can use the PA model to map their processes in a manner that is understandable to the customers and to those who represent the customers — third- and second-party auditors. Plus there is no wasted time in doing so; that is, no other conversion, translation or delineation is necessary. Once mapped (and maintained for accuracy), the requirements of ISO 9001:2000 for "sequence and interactions" are met along with nearly every other system requirement. How the system is managed and how it meets customer expectations and requirements cannot be satisfied by the PA model alone, obviously. But because the system is visible, integrated, focused on the customer (internal and external) and more amenable to planned arrangements, it is a huge step in the right direction. It cannot force management commitment, for instance, but it can make it harder to deny the obvious.

ISO 9001:2000 Applied to the PA Model

ISO 9001:2000 requires that each organization define its process management approach. While no specific approach is required, you need a good starting point to analyze your functions in terms of the process approach. Try to view the system from the point of view of your customer, what can be termed customer-oriented processes (COPs). Identification and application of COPs sets up a fairly easy process approach hierarchy to understand, implement and follow. COPs are those processes that interface with external customers. Support processes, the second level, are those processes that support or feed the COPs. Management processes are those processes that keep the COPs and support processes running smoothly.

Can an organization set up its quality management system with COPs, support processes and management processes? Chances are they already have; they simply have not acknowledged it. They have COPs or they would not be in business. They have some type of support processes network to support what they do with their customers. And they must make decisions, collect data, make changes, etc. — all of which are actions that are required to manage their processes and their overall system. Other logical ways to think about how an organization works may be just as valid.

The system your organization defines should not only work but also be visible and complete. Be prepared to explain and sometimes defend your system to people who do business with the organization, especially if that customer is an OEM or one of its emissaries (third- and second-party auditors).

10 Customer-Oriented Processes

The 10 generic COPs supported by most organizations in one form or another are:

1. Market Analysis/Customer Requirements.

2. Bid/Tender.

3. Order/Request.

4. Product and Process Design.

5. Product and Process Verification/Validation.

6. Product Production/Manufacturing.

7. Delivery.

8. Payment.

9. Warranty/Service.

10. Post Sales/Customer Feedback.

The preceding processes need not be formally recognized by the organization. They certainly can be, but the main point is that these processes represent typical interfaces with them. Each, of course, may lead to customer satisfaction or dissatisfaction.

Product Realization – A Focus

It stands to reason that many customers conclude that items four through seven of the 10 generic COPs are most important and that number six is probably the most important of all.

If you have any doubts about this, simply examine the checks, mechanisms, tools and/or methodologies that each of the customers have in place to make sure the parts/products/services they receive are good parts/products/services. They tell the real story.

What does this mean to an organization? It provides focus. An organization can't stop supporting the other COPs, but it can make sure that they are aligned with those that are most important to their customers.

If the organization is providing the service of auditing (a registrar), then those COPs should take the highest priority. But once again, it is a matter of focus. If four through seven are audited well, the other COPs will most certainly be covered simply because they can't be avoided. So, the message seems to be, find a part and follow it up and down the manufacturing stream.

COPs Mapped

In examples that follow in this section the 10 generic COPs are mapped (though not completely — because the application is nonspecific not all sections can be completed) via the PA model. What will be included are some suggested and generic support and management processes, as well as some completion of applicable measurements and requirements. Keep in mind these are only suggestions.

Following the example COPs on the PA model will be a look at the organization's system from a customer perspective. This is no different from what has already been discussed; it simply puts the emphasis in a more visual manner in the context of COPs.

Overall System Structure

To help you visualize the overall structure, the graphic on the following page summarizes (1) the overall model identification, (2) identification of COPs, (3) analysis of COPs with the PA model and finally (4) the recording of COPs on PA models from information obtained in the analysis, as well as internal and external requirements/analysis.

A Way To Think About It

Perhaps a good way to think about the structure and relationship of this graphic is to imagine a computer software program that allows you to build a virtual structure. The first structure is a circle, which symbolizes your organization. Loops are added as the COPs are identified and defined. Next, the PA model is used to see if the COPs have been considered and all risks have been mitigated. As a consequence of the analysis, and to record the findings, the PA model is fully used. The PA model provides direct assistance to auditing, record keeping, identification of related processes, requirements, measurements, etc. and provides a ready means to communicate the structure and operation of the system to stakeholders. Not shown in the graphic, but which would also be a

Figure 24 — Total Organizational System and Process Analysis

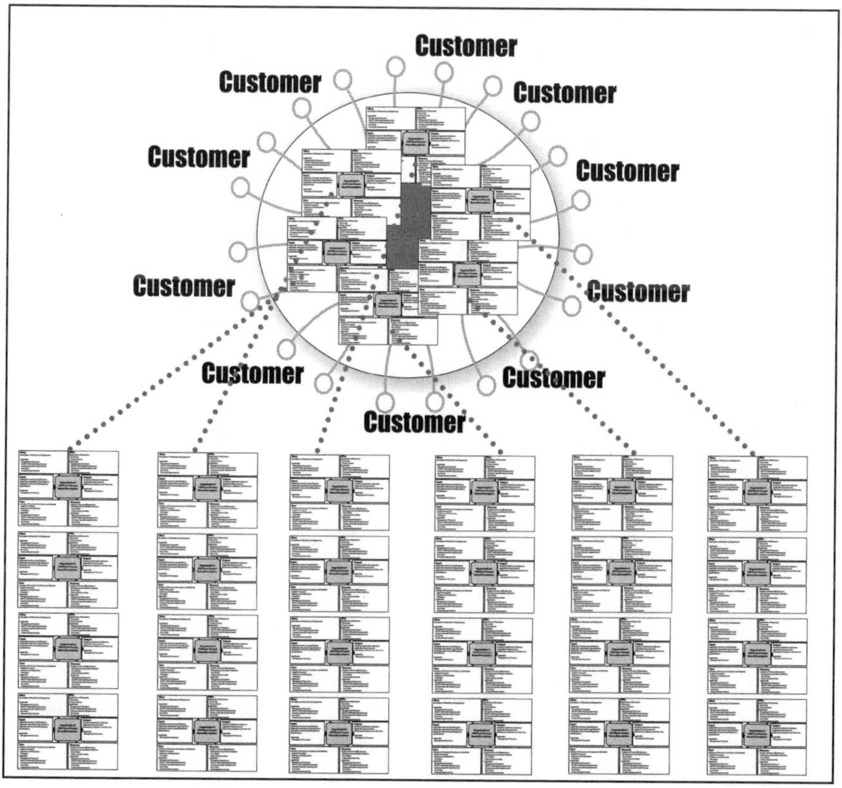

part of this virtual system, would be PA models completed for each support and management process identified in the COP/core PA model (there are a total of 10 COP/core PA models in this graphic to reflect 10 common COPs). The PA model is intended to symbolize the defined processes supporting each COP. The number then, is not a magic one — there could be more or less. The graphic is provided to help you visualize the larger system and how it is connected.

Mapping COPs

The PA model on page 107 uses a generic COP to illustrate the completion of a PA model. The details of the input, output, who, what, how and measure are provided generically. The intent is to give a person using this tool some insights into what may be placed in the boxes and how that may help the organization or group to which that individual belongs. The details provided are generic; therefore, some of them will fit your situation, while others may not. Like most things related to the process approach and management system implementation, as much flexibility as possible is being provided within a framework of a generic system, in order that uniqueness of the organization is embraced and that the final system will work for all members of the organization, not just auditors and quality management system personnel.

Successive PA models must be completed for COP/core-level/top-level processes and for each support process. Once completed, your quality management system (and business system) will be visible and on its way toward a process approach management system.

Suggestion: For ease of administration and access, it will be quite helpful if the system captured on PA models is placed on an electronic platform, particularly a data-based platform.

Process Approach Model

Organization:

Process Owner:

Completed by:

Date Completed:

What *(Resources)*

Description of Materials and Equipment: Office equipment, computers, software, communication equipment

Applicable

Management Processes: Management review, resource management

ISO/TS 16949:2002 Requirements: Clauses 4 and 5 have the most direct application

Customer-Specific Requirements: Per contract

Core Tools: APQP, FMEA

Continual Improvement: As applicable

Who *(Management Responsibility)*

Application of Resources: Training in contract review

Personnel: Sales, customer rep., management, engineering, quality

Process Owner: Management, sales or engineering

Applicable

Management Processes: Resource management, management review, internal audit

ISO/TS 16949:2002 Requirements: Clauses 4 and 5 have most direct application

Customer-Specific Requirements: Linked to contract

Core Tools: APQP

Continual Improvement: As applicable

Input

Applicable Customer Specifications: Per product or service needed

Applicable Organization Specifications: Management review, resource management

Applicable Governmental/Regulatory Specifications: As applicable to product or service

Applicable

Management Processes: Management review

Organization's COP/Core Process Name/Description: *Order/Request*

Output

Customer Satisfaction Indicators: A signed contract

Applicable Measurements: Dollar amounts, order size, concessions/deviations

Performance Indicators and Their Use: Training, changes, deviations, cost overruns = shape of future bids

Applicable

Management Processes: Management review

How *(Product Realization)*

Linkages of Processes, Procedures and Methods: Linkage to contract

Support Processes: Contract review, design review (preliminary), management review, training, design objectives, special characteristics, preliminary bill of materials

Sequence and Interaction: Prior to design

Procedures: As applicable — contract review required

Applicable

Management Processes: Resource management, management review

ISO/TS 16949:2002 Requirements: Clause 4 and 5 most applicable

Customer-Specific Requirements: Contract requirements

Core Tools: APQP, FMEA (preliminary)

Continual Improvement: As applicable

Measure *(Analysis and Improvement)*

Analysis of Process Effectiveness: Bid to contract ratio

Measurement of Metric Selection: Bid to contract ratio, time to review quotes, number of changes to quote, timing

Use of Data: Decision — resources, decisions (capability and capacity), adequacy of review process, capability of review teams

Actions Based on Data: Resource applications, methods changes, process improvements

Applicable

Management Processes: Management review

ISO/TS 16949:2002 Requirements: Clause 8 is most applicable

Customer-Specific Requirements: Often N/A

Core Tools: SPC

Continual Improvement: As applicable

Aligning the Process Approach with First-, Second-, Third-Party Audits

Once the system is running, the audit function, including first-, second- and third-parties, can employ the same tools to audit the system and processes. Auditors use the PA model to test the process, as identified for fundamental considerations of identification, risk, linkages and adequacy. Once that has been accomplished, the PA model can be used as an audit road map by the auditor to check for linkages to other processes, specifications, requirements, compliance to standards, measurement, testing and management.

Figure 25 depicts the analysis, structuring and construction.

Figure 26 puts the whole process together in one concise package. This chart is a summary look at the entire process approach implementation, analysis and confirmation.

And there you have it. Hope your improvement journey is a good one.

Figure 25 — Total Organizational System and Process Analysis

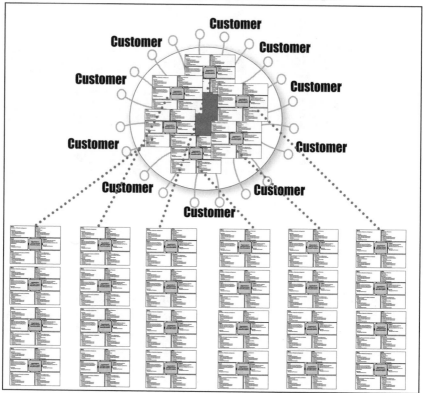

Figure 26 — Five Steps to Process Approach Implementation

Step 1

To analyze the organization's system "find the processes"

Components: 1) Entire organization, 2) PA model.

Actions: Identify all of the main processes without classifying them.

Deliverables: Listing of all main processes.

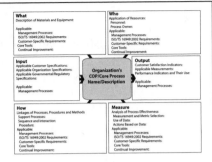

Step 2

To define the system's model based on work from Step 1

Components: A single concept model (in this case the Octopus, which is based on the COP). *Note: two other models are shown, the value stream model and the APQP model.*

Actions: Determine model.

Deliverables: A model that identifies/classifies key processes, sequence and interaction.

Octopus model　　*APQP model*

Value stream model

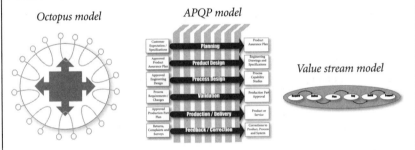

Step 3

To define the support and management processes

Components: 1) The actual processes, 2) PA model, 3) Levels.

Actions: Identify the support and management processes.

Deliverables: Listing of all support and management processes to the detail appropriate to the organization.

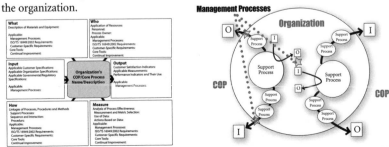

Figure 26 — Five Steps to Process Approach Implementation *(continued)*

Step 4

To adequately define and communicate the QMS

<u>Components:</u> 1) System documentation, 2) PA model.
<u>Actions:</u> Develop the documentation that can adequately communicate the system.
<u>Deliverables:</u> Documents, tools, procedures and forms.

Quality Manual

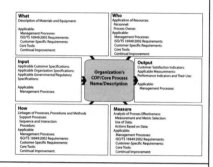

Step 5

To align first- and third-party audits by standardizing the analytic and communication tools

<u>Components:</u> 1) PA model for system, 2) PA model for internal audit, 3) PA model for third-party audit.
<u>Actions:</u> Communicate system to internal and external auditors.
<u>Deliverables:</u> 1) Picture of the system, 2) Internal audit, 3) External audit, 4) Accurate assessment of organization.

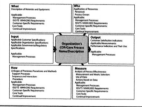

Software Support for the Process Approach to Management

The process approach contains within it a kind of double-edged logic; that is, it provides visibility and it requires visibility to be effective.

You may be saying to yourself that really doesn't seem to be a problem; after all, visibility is a good thing. But let's think it through to a more detailed level and imagine making a practical application of the approach within an organization. You will quickly find that some things become less visible because there are so many things that are now visible. It's kind of like a variation of an old saying "You can't see the forest for the trees."

For the visibility gained through the process approach to have value for the organization, it must be of value to the people in the system. A management system is, after all, the result of people carrying out actions that result in intended outcomes, products, etc. But when you think about it, it is a rare manager, indeed, who can keep the picture of a complex system in his or her head. Most people are able to retain the general outlines of systems and can even retain some of the more detailed processes; but it is a rare skill for anyone to be able to reconstruct a detailed model of a complete system in his or her mind, even if the details of the system are visible to the individual.

So, what is needed to take advantage of the visibility gained via a detailed analysis and implementation of the process approach? This is an ideal situation for a software application.

We reviewed a number of software applications that claimed to support the process approach to management. Most fell short of expectations. They did so for a variety of reasons. Many software applications employed flowcharting to capture the sequence and interaction of processes. The problem with this approach is that processes and their interactions within an organization operate in three dimensions. Two-dimensional drawings don't provide adequate visibility of the true nature of the system or process. Other software provided too much data in conjunction with the picture/graph/illustration, which not only reduced the visibility of the system or process, but it actually discouraged us from spending any length of time looking at the collection of data.

Terrapene Process Management Suite

As a result of these shortcomings, my organization developed the Terrapene Process Management Suite™ (Terrapene). It is better than the rest in my analysis of software packages.

Essentially Terrapene provides an effective platform for the three fundamental areas of process-based management:
+ Organization/Implementation (Visibility).
+ Operation/Management (Communication).
+ Improvement/Auditing (Measurement/Analysis).

Figure 27 — 10 Top-Level/Core Processes

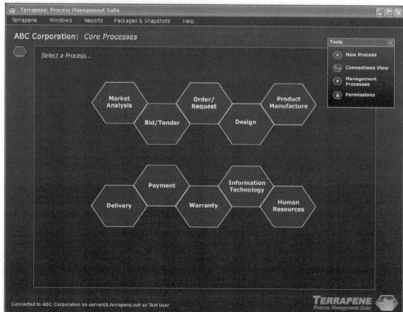

Figure 28 — Analyzation of One Top-Level/Core Process

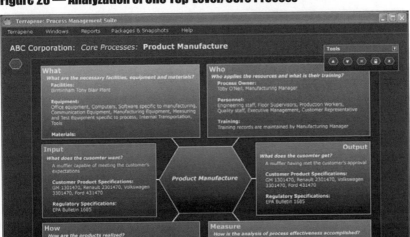

To ensure that the visibility, communication and analysis are useful and not overwhelming, Terrapene employs as its basic tool something called a "turtle" diagram (hence the name Terrapene = box turtle). The turtle allows the user to divide a process into six distinct areas — Input, Output, What, How, Who and Measure. Terrapene displays these divisions on screen supported by questions to facilitate completion and use. When a turtle is completed, Terrapene is able to display the information in a manner that is most useful to the user. In other words, the user can look at as much or as little as they want at any one time, and can navigate as deeply into the system as needed at the time.

For example, an organization has identified that it has 10 top-level/core processes. On the Terrapene software those processes look like Figure 27.

Let's assume an organization would like to analyze one of its top-level/core processes, Product Manufacture. On Terrapene, the screen might look like Figure 28, upon completion.

Now, let's assume the organization would like to examine one aspect of the process in more detail, including attaching and/or finding pertinent documents and comments or notes. On Terrapene this might look like Figure 29, depending on the numbers and length of documents and notes.

Figure 29 — Detailed View of a Single Aspect

As of this printing, Cadillac Products Automotive Company of Troy, Michigan, reports a $1.1 million savings at their Rogers City, Michigan, plant in the first year of Terrapene use. Cadillac attributes the $1.1 million savings directly to its use of Terrapene.

While it is not necessary to purchase software to implement the process approach, it is important that you purchase good software if you intend to go this route. There is little doubt that good software would be tremendously useful in the implementation, execution and maintenance of your system.

Index